# *Mud, Blood & Cold Steel*

The Retreat From Nashville, December 1864

## Mark Zimmerman
Zimco Publications LLC

**Also by Mark Zimmerman
and available from Zimco Publications LLC:**

*Iron Maidens and the Devil's Daughters:
U.S. Navy Gunboats versus Confederate Gunners and Cavalry
on the Tennessee and Cumberland Rivers, 1861-65*

*Guide to Civil War Nashville, 2nd Edition*

*God, Guns, Guitars & Whiskey:
An Illustrated Guide to Historic Nashville, Tennessee, 2nd Edition*

*Gone Under:
Historic Cemeteries and Burial Grounds
of Nashville, Tennessee, 2nd Edition*

# *Mud, Blood & Cold Steel*

The Retreat From Nashville, December 1864

## Mark Zimmerman
Zimco Publications LLC

Mud, Blood & Cold Steel: The Retreat From Nashville, December 1864
Copyright © 2020 Mark Zimmerman
Zimco Publications LLC
Website: zimcopubs.com
Email: info@zimcopubs.com

ISBN   Paperback   978-0-9858692-6-7
ISBN   eBook       978-0-9858692-7-4

All rights reserved.

All Text, Photographs, and Maps by the Author Unless Otherwise Noted

Publication Contributor: James D. Kay, Jr.

No part of this book may be reproduced or transmitted in any form or by any means — electronic or manual, including photocopy, scanner, email, CD or other information storage and retreival system — without written permission from the author, except for personal use or as provided to the news media and book sellers.

Printed in the United States of America.

The text in this book, which is published to inform and entertain, should be used for general information and not as the ultimate source of educational or travel information. Every effort has been made to ensure the accuracy and relevence of information in this book but the author and publisher do not assume responsibility for any errors, inaccuracies, omissions or inconsistencies within. Any slights of people, places, or organizations are strictly unintentional.

The names of organizations and destinations mentioned in this book may be trade names or trademarks of their owners. The author and publisher disclaims any connection with, sponsorship by or endorsement of such owners.

Content related to this publication can be found on the publisher's website at: zimcopubs.com

*Dedicated to the current and past members of the Battle of Nashville Trust, formerly the Battle of Nashville Preservation Society, who have worked tirelessly to preserve what remains of the battlefield — hallowed ground.*

## Acknowledgements

*The author would like to acknowledge the valuable assistance of Jim Kay, president of the Battle of Nashville Trust, who read the manuscript, offered significant clarifications, and guided me on a tour of the battle grounds. Sam Huffman, board member of Save the Franklin Battlefield, was kind enough to guide me at the site of the West Harpeth battle and discuss other matters pertaining to Franklin. Ross Massey contributed to the two Battle of Nashville maps.*

## Table of Contents

Foreword ............................................................................................... 3
   *Map: Hood's Retreat* ...................................................................... 4
   *Timeline of Events* ......................................................................... 5
   *Major Players* ................................................................................ 6

**Introduction** ...................................................................................... 11

**Prelude** ............................................................................................. 15
   *Map: Hood's Tennessee Campaign* ............................................ 20

**The Battle of Nashville Begins** – Thurs., Dec. 15th, 1864 ............ 29
   *Map: Compton's Hill* ................................................................... 44

**Compton's Hill** – Fri., Dec. 16th, 1864 ............................................ 45
   *Map: Compton's Hill to Brentwood* ............................................ 54
   *Map: Peach Orchard Hill* ............................................................. 58
   *Map: Battle at the Barricade* ....................................................... 66
   *Map: Pvt. Johnston's Recollection* ............................................... 67
   *Map: Brentwood to West Harpeth* .............................................. 76

**Hell-Bent for Leather** – Sat., Dec. 17th, 1864 ................................ 77
   *Map: Battle at Hollow Tree Gap* ................................................. 80
   *Map: Battle at the West Harpeth* ................................................ 88
   *Map: Spring Hill to Columbia* .................................................... 98

**Forrest Rejoins Hood** – Sun., Dec. 18th, 1864 .............................. 99

**Stymied by Rising Waters** – Mon., Dec. 19th, 1864 ................... 105
   *Map: Columbia to Pulaski* ......................................................... 110

**The Rearguard Reorganizes** – Tues., Dec. 20th, 1864 ................ 111

**Desperate Desolation** – Wed., Dec. 21st, 1864 ............................ 119

**Over The Duck** – Thurs., Dec. 22nd, 1864 .................................. 123

**Blunting The Pursuit** – Fri., Dec. 23rd, 1864 .............................. 127
   *Map: Battle at Richland Creek* .................................................. 130

**Richland Creek** – Sat., Dec. 24th, 1864 ........................................ 131
   *Map: Pulaski to the Tennessee River* ......................................... 134

**Christmas Day** – Sun., Dec. 25th, 1864 ....................................... 139
   *Map: Battle at Anthony's Hill* ................................................... 144

**Finale** – Mon.-Thurs., Dec. 26th-29th, 1864 ................................ 149
   *Map: Battle at Sugar Creek* ........................................................ 150

**Aftermath** ....................................................................................... 159
   *Shy's Hill (Compton's Hill) Today* ............................................ 166
   *Orders of Battle* .......................................................................... 168
   *Bibliography* ............................................................................... 180

## Foreword

I am so humbled that my friend, Mark Zimmerman, would ask me to contribute to this fine work. So many books have been written about the famous Battle of Nashville, but most have missed the details of the retreat down Franklin Pike, through the gap at present-day Radnor Lake, and the famous Battle of the Barricade on Granny White Pike that closed the fighting on December 16, 1864. Mark has been able to compile a concise overview of all the action, which is a very difficult task.

I have walked the Nashville battlefield since 1966, when I was six years old, living on land where history was created long ago. The Nashville battlefield has essentially been obliterated. There are only a few people left who know exactly where the troops fought and died. We know where the little spots of history — which have not been wrecked by development — still stand. The relics continue to come from the dark topsoil, revealing death, destruction, and a desperate fight for this country. Mark has seen many of these small treasures with me and has gone to great efforts to map some of these sites.

In the preservation world, there are only a handful of people who step up to make an enduring financial contribution. Long ago, the author donated all the proceeds from his *Guide to Civil War Nashville-1st Edition* to the Battle of Nashville Trust. The sale of these books — every year — has been an incredible gift to the Trust to maintain its protected sites in Nashville. The Battle of Nashville Trust can never express in words its gratitude to the author for this gift. We are so grateful that he has another work that is being published that will give those with an interest in the Nashville campaign and the retreat another detailed and balanced perspective.

On behalf of a grateful Battle of Nashville Trust, I thank the author for his hard work, and I know that you will enjoy this.

James D. Kay, Jr.
Battle of Nashville Trust, Inc.
01-May-2020

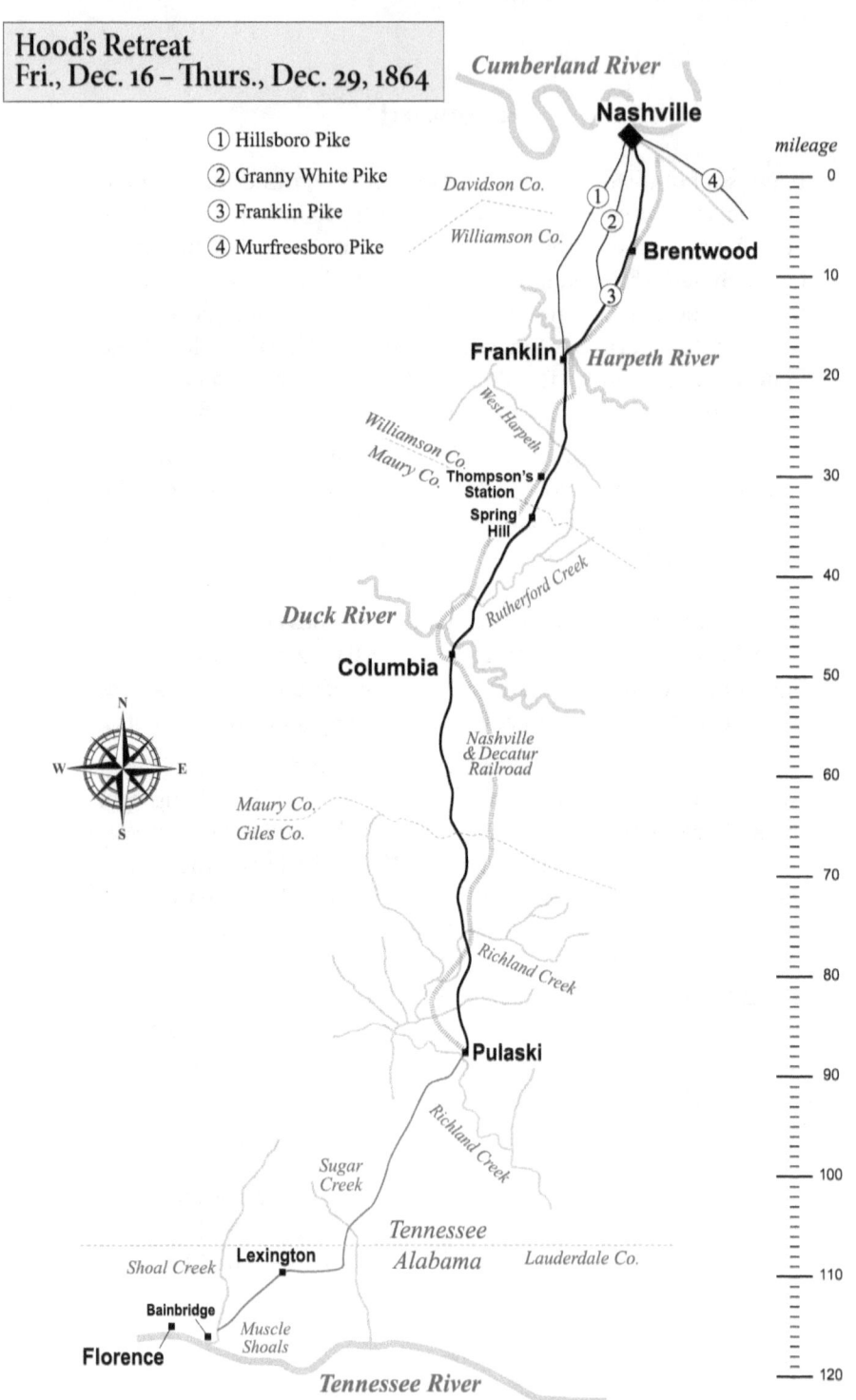

## Timeline of Events - 1864

| | |
|---|---|
| Nov. 13-20 | Army of Tennessee crosses Tennessee River |
| Nov. 15 | Sherman begins March to Sea |
| Nov. 29 | Affair at Spring Hill |
| Nov. 30 | Battle of Franklin |
| Dec. 7 | Battle of Cedars at Murfreesboro |
| Dec. 15 | First day of Battle of Nashville |
| Dec. 16 | Compton's Hill, Battle at the Barricade |
| Dec. 17 | Hollow Tree Gap, Harpeth River, West Harpeth |
| Dec. 18 | Spring Hill |
| Dec. 19 | Rutherford Creek |
| Dec. 20 | Columbia |
| Dec. 23 | Warfield's |
| Dec. 24 | Lynnville, Richland Creek |
| Dec. 25 | Anthony's Hill |
| Dec. 26 | Sugar Creek |
| Dec. 27 | Hood's army back across Tennessee River |
| Dec. 29 | Thomas officially calls off pursuit |

**Major Players**

### Federal

Col. Datus E. Coon, 34, of New York, cavalry brigade commander under Hatch

Brig. Gen. John T. Croxton, 28, of Kentucky, cavalry brigade commander under Wilson

Brevet Brig. Gen. John H. Hammond, 31, of New York City, cavalry brigade commander under Knipe

Col. Thomas J. Harrison, 40, of Indiana, cavalry brigade commander under Johnson

Brig. Gen. Edward Hatch, 31, of Maine and Iowa, 5th cavalry division commander under Wilson

Lt. Joseph S. Hedges, 28, of Ohio, commander of Wilson's escort, the 4th US Cavalry, and recipient of the Congressional Medal of Honor

Col. Lucius F. Hubbard, 28, of Minnesota, brigade commander who stormed Compton's Hill

Brig. Gen. Richard W. Johnson, 37, of Kentucky, 6th cavalry division commander under Wilson

Brig. Gen. Joseph Farmer Knipe, 41, of Pennsylvania, 7th cavalry division commander under Wilson

Col. William R. Marshall, 39, of Minnesota, brigade commander who stormed Compton's Hill

Brig. Gen. John McArthur, 38, of Chicago but native of Scotland, commander of division that stormed Compton's Hill

Col. William L. McMillen, 33, of Ohio, brigade leader who led charge up Compton's Hill

Major Gen. John McAlister Schofield, 33, of New York, commander of XXIII Army Corps

Major Gen. Andrew Jackson Smith, 49, of Pennsylvania, commander of XVI Army Corps

Col. George Spalding, 28, of Michigan, commander of the 12th Tennessee (US) Cavalry, victor of duel at Battle at the Barricade

Major Gen. George Henry Thomas, 48, of Virginia, overall commander of Federal forces at Nashville

Major Gen. James Harrison Wilson, 27, of Illinois, young, aggressive commander of Federal cavalry forces

Brig. Gen. Thomas John Wood, 41, of Kentucky, commander of IV Army Corps, often leader of pursuit by infantry

## Confederate

Brig. Gen. Frank Armstrong, 29, cavalry brigade leader under Red Jackson

Major Gen. William Brimage Bate, 38, of Tennessee, division leader at Compton's Hill

Col. Tyree Harris Bell, 49, of Tennessee, cavalry brigade leader under Buford

Brig. Gen. Abraham Buford, 44, of Kentucky, cavalry division leader under Forrest; wounded at Buford's Station

Brig. Gen. James Ronald Chalmers, 33, of Mississippi, cavalry division leader under Forrest, commander at Nashville

Major Gen. Benjamin Franklin Cheatham, 44, of Tennessee, corps commander under Hood

Major Gen. Henry D. Clayton, 37, of Georgia, division commander under Lee

Col. David Coleman, 40, of North Carolina, commander of Ector's Brigade, a fire brigade

Major Gen. Nathan Bedford Forrest, 43, of Tennessee, overall cavalry commander, joined retreat at Columbia and took command of rearguard

Capt. William D. Gale, Assistant Adjutant General, Stewart's Corps, keen observer of operations

Brig. Gen. Randall Lee Gibson, 32, of Louisiana, brigade leader in Clayton's Division, rearguard action at Franklin

Brig. Gen. James T. Holtzclaw, 31, of Georgia, brigade leader in Clayton's Division, action on Franklin Pike

Lt. Gen. John Bell Hood, 33, of Kentucky and Texas, overall Confederate commander at Nashville

Brig. Gen. Henry Rootes Jackson, 44, of Georgia, brigade commander under Bate, captured at Compton's Hill

Brig. Gen. William Hicks "Red" Jackson, 29, of Tennessee, cavalry division commander under Forrest

Lt. Col. David Campbell Kelley, 33, of Tennessee, battalion leader under Rucker

Lt. Gen. Stephen Dill Lee, 31, of South Carolina, corps commander under Hood, wounded at Franklin

Capt. John Morton, 22, of Tennessee, chief of artillery under Forrest

Brig. Gen. Daniel H. Reynolds, 32, of Arkansas, brigade commander under Walthall, rearguard action at Compton's Hill

Brig. Gen. Lawrence Sullivan "Sul" Ross, 26, of Texas, cavalry brigade commander under Jackson

Col. Edmund W. Rucker, 29, of Tennessee, brigade commander under Chalmers, wounded and captured at Battle at the Barricade

Pvt. Philip Daingerfield Stephenson of the Fifth Washington Artillery Co. of New Orleans, wrote vivid memoir of his wartime experiences

Major Gen. Carter Littlepage Stevenson Jr., 47, of Virginia, division commander under Lee, leader of rearguard until Franklin

Lt. Gen. Alexander Peter "A.P." Stewart, 43, of Tennessee, corps commander under Hood

Major Gen. Edward Cary Walthall, 33, of Mississippi, division commander under Stewart, leader of infantry rearguard after Columbia

Pvt. Sam Watkins, 25, of Co. H, 1st Tennessee, wrote vivid memoir of his wartime experiences

## Introduction

The Battle of Nashville, fought December 15-16, 1864 just south of Tennessee's capital, is considered one of the most decisive battles of the Civil War. Decisive because it ended the last major Confederate offensive in the Western Theater, and decisive because it was an undisputed victory for the Federals, ending in a rout. The Confederate Army of Tennessee under Major General John Bell Hood was attacked by the Federals under Major General George Henry Thomas and soundly defeated. America's preeminent military historian, the late Russell F. Weigley, reaffirmed that "Nashville ranks as probably the most complete battlefield victory of the war." Nashvillian Stanley Horn's definitive 1956 book on the subject is titled *The Decisive Battle of Nashville*. But Hood's army was not destroyed, as some historians and authors contend. The rearguard of Hood's army fought valiantly for the next ten days, under continuous heavy pursuit and in miserable winter weather conditions. Federal forces spearheaded by the troopers of Major General James Harrison Wilson chased Hood's routed ragtag army more than one hundred miles south to the Tennessee River in northern Alabama. Given their overwhelming numbers and firepower, and the beaten-down condition of Hood's army, the Federals under Thomas and Wilson should have rather easily bagged the entire rebel army and either annihilated them or shipped them off to Northern prisoner-of-war camps. The fact that neither one of those possibilities occurred is testament to Yankee overconfidence and floundering, the appalling winter weather, the rugged terrain, and, last but not least, the tactics of the Confederate rearguard.

"There was no pursuit and no rearguard action during the entire war to compare with that during Hood's retreat," declared historian Paul H. Stockdale, a viewpoint echoed by renowned historian

Edwin Bearss. Cavalry expert Edward G. Longacre noted, "The pursuit was one of the most devastating in American history." Ross Massey, historian for the Battle of Nashville Trust, said, "No army in the war endured a more miserable and depressing episode than did the Army of Tennessee on this retreat."

There are few examples of successful pursuits following battlefield victories in any theater of the Civil War. Failure to act decisively was one reason, as at Shiloh, Antietam, and Gettysburg. But even when pursuit was employed, as at Chickamauga and Missionary Ridge, seldom does the victor destroy the defeated force. Christopher Einolf, a biographer of Thomas, noted one of the main reasons—a retreating army falls back on its own supply line, while a pursuing force tends to outrun its own supplies.

Horn elaborated on the aftermath of the two-day battle at Nashville: "The next 10 days were a nightmare of nerve-wracking hardship and struggle to both armies. Alternating marching and fighting, worn down by battle fatigue and sheer physical exhaustion, they somehow managed to carry on an almost continuous running battle from Nashville to the Tennessee River. The weather was abominable—rain, sleet, and snow, with below freezing temperatures. The wagons and guns churned the roads into seemingly bottomless quagmires, which froze into sharp edged ruts during the cold nights. The heavy rains not only drenched the suffering soldiers but soon flooded the streams and made the passage of each of them a serious problem."

Popular historian Hampton Sides was writing about Korea 1950, but he could have been referring to Middle Tennessee in December 1864: "Whatever euphemism one wanted to use, all the martial textbooks agreed on this point: Even under more favorable circumstances, a disciplined, well-choreographed fighting withdrawal was one of the trickiest maneuvers in military science … It was hard enough for an army to defend itself when dug in; to do so while on the move, with a numerically superior army attacking every inch of a rearward march, was next to impossible. Yet some battlefield situations offered only one solution beyond surrender or destruction—and that solution was a swift exit."

General Thomas would have known this maxim well; he was called the Rock of Chickamauga for holding fast at that bloody 1863 battle, the largest of the Western Theater, and then retreating from the enemy at nightfall to the safety of Chattanooga. For his victory a year later, an offensive masterpiece, Thomas became known as the Sledge of Nashville.

The running pursuit began late on the afternoon of the second day of battle when anxious Federal infantry stormed the heights of Compton's Hill (now known as Shy's Hill) and the Confederate soldiers turned and fled, initiating a full-fledged rout. With the Federals and Wilson's troopers nearly surrounding the hill and cutting off the closest escape route, the Confederate soldiers skirted the Overton Hills and fled southeastward to the Franklin turnpike. Units of Hood's army organized to establish holding actions which allowed much of the infantry to escape through Brentwood and surge southward to the village of Franklin. Fighting was fierce on December 17th, with hand-to-hand combat north of Franklin and a major confrontation at a tributary of the West Harpeth River. The command of the Confederate rearguard changed hands several times in the first few days as Federal troopers, armed with repeating rifles, tried to outflank the rebels on the turnpike. Acres of knee-deep mud sapped the strength of man and beast. The use of the macadamized turnpike, which ran all the way from Nashville to Franklin to Columbia to Pulaski, was vital to both sides during the retreat.

Many rivers and streams had to be forded. Destroyed bridges had to be replaced. Rainy conditions turned small creeks into raging torrents; there was much skirmishing along Rutherford Creek, which was difficult to cross. The arrival of the Federal pontoon train was delayed due to critical errors. Then the weather turned bitterly cold, with snowfall. The Duck River at Columbia was a major obstacle to the Federals, as was the arrival of Confederate Major General Nathan Bedford Forrest. The wily general employed delaying tactics to hasten the retreat while battling his tormentors all the way through Pulaski to the state line. Major conflicts were fought at Lynnville and Richland Creek north of Pulaski, and at Anthony's Hill and Sugar Creek south of that town. By that time, the Confederate troops were slowly crossing the Tennessee River to safety on a rickety pontoon bridge as the last-chance fighting evolved into a purely cavalry match.

Back on December 16th, the Confederates atop Compton's Hill had several options — they could desert, surrender, fight to the death, or flee. That their fate was not decided for the next ten days was testament to the willpower of both blue and gray.

As Winston Churchill, an avid scholar of American history and the Civil War, said, in emphasizing the fighting spirit of Americans, the Civil War was fought "to the last desperate inch."

## Prelude

John Bell Hood's military campaign and the overall strategic situation for the Confederates in November-December 1864 can be summed up in one word—desperate.

To gain the proper perspective on Hood's plan to invade Tennessee one must go back to the summer of 1864 and the predicament of the Confederate army defending Atlanta, Georgia. On July 18th, with Federal General William Tecumseh Sherman's three armies at the city's outskirts, Confederate President Jefferson Davis relieved the current commander, Joseph Johnston, and gave Hood command of the Confederate Army of Tennessee. Known throughout the war for his bravery and aggressiveness, Hood immediately launched three fierce attacks against Sherman's forces, none of them successful and all costly in casualties, especially within the officer corps. Military observers note that Hood effectively crippled his army at Atlanta, losing 13,000 casualties. He abandoned Atlanta to Sherman on September 2nd.

Some historians contend that Hood should never have been given army command in the first place. Richard McMurry, chronicler of the Army of Tennessee, noted: "It was Hood's tragedy that he was such an excellent soldier, but such a poor general."

Hood had earned glory early in the war at Gaines Mill, Virginia, and Sharpsburg, Maryland, but he lost the use of his left arm due to a wound at Gettysburg. Later, his right leg had to be amputated just below the hip due to a wound at Chickamauga. Most victims did not survive such a leg wound and amputation — Hood's detached leg was transported with him so that it could be buried with his body. But Hood survived, and his leg was buried by itself, reportedly near Tunnel Hill, Ga. His arm in a sling, fitted with an artificial leg, he used crutches to walk, and three men were required to hoist and strap him into the saddle. The wounds and setbacks served only to motivate Hood to regain the glory that now eluded him. Nobody doubted Hood's bravery and courage, but colleagues, including Robert E. Lee, wondered if he possessed the intellect and patience to lead an army.

Born in Kentucky in 1831, Hood had served before the war in the U.S. cavalry in Texas along with George Thomas, and had led a punitive expedition against the Comanches, engaging battle with a revolver in one hand, reins and sawed-off shotgun in the other. In a brisk skirmish with the Indians, an arrow pinned Hood's hand to his saddle as his men eventually drove off the band. When the Civil War broke out, Hood led a brigade of tough Texans into battle, earning laurels for his attack up a hillside at Gaines Mill and his persistence at the bloody cornfield at Sharpsburg (his brigade sustaining a casualty rate of 86 percent). Following his wounding at Gettysburg, he became the toast of Richmond during his convalescence, befriending Jefferson Davis and becoming engaged to a coquettish socialite. Hood was then transferred to the Western Theater along with Longstreet's troops and was grievously wounded at Chickamauga. Later, in 1864, during Sherman's campaign to Atlanta, Hood intrigued against his commander, Joseph Johnston, urging Davis to put him in charge so the army could take the offensive.

On September 27th, at a train stop in Macon, Ga., President Davis told a crowd: "Our cause is not lost. Sherman cannot keep up his long line of communication; and retreat sooner or later he must. And when that day comes, the fate that befell the army of the French Empire in its retreat from Moscow will be reenacted." Little did Davis realize that the fate he forecast for Sherman would befall Hood instead.

Hood abandoned Atlanta and headed northward to attack Sherman's line of communication (supply line). In mid-November, Sherman took the elite of his armies and headed to Savannah and the sea, while Hood moved into northern Alabama to stage his Tennessee invasion. Middle Tennessee and Nashville, the state capital, had been captured by the Federals in 1862 and held ever since. Nashville was converted into a huge supply depot and transportation hub, with a gigantic earthen fort and supply depot built outside nearby Murfreesboro, named Fortress Rosecrans, so large it encompassed sections of Stones River, the Nashville & Chattanooga Railroad, and the Nashville turnpike. The railroad was protected from Confederate cavalry raiders such as Bedford Forrest by U.S. Colored Troops (commanded by white officers) garrisoned in several blockhouses along its length from Murfreesboro to Nashville.

As Hood turned north to cut Sherman's railroad supply lines, a hard blow landed on November 8th. Abraham Lincoln won re-election as U.S. President, due in large part to the Federal capture of Atlanta, thus ensuring the continued prosecution of the war. Shortly thereafter, Sherman received approval of his bold plan to abandon his supply lines and march his army through Georgia to the sea. In setting off, Sherman sent Thomas, Wilson, and Major General John McAlister Schofield back to Middle Tennessee to hold that region, including Nashville. About that time, Hood set off on his plan to invade Tennessee, recapture Nashville, and then terrorize Yankees north of the Ohio River. That or turn east to reinforce Lee's besieged troops at Richmond-Petersburg, Virginia.

One of the knocks against Hood was his lack of attention or interest in logistics. Hood was forced to wait at Tuscumbia, Alabama, for three weeks while his commissary officers attempted to accumulate 20 days worth of rations for the upcoming campaign. This was complicated by the lack of viable rail facilities. He also needed to wait for the arrival of Forrest, off on a raid. Finally, on November 21st, the campaign got underway. Separate routes were taken northward by Hood's three corps commanders:

- Major General Benjamin Franklin "Frank" Cheatham, 44, a native Nashvillian, a hard fighter, a hard drinker, and beloved by his men. He served well in the Mexican War and fought in nearly every battle of the Army of Tennessee. He was elevated to corps commander at the beginning of the campaign, replacing William Hardee.

- Lieutenant General Alexander Peter "A.P." Stewart, 43, born in Tennessee, attended West Point, and then taught as a professor at Nashville University. Known as Old Straight, he was a honorable man of manners who fought competently and without drama, a true Southern gentleman. He assumed corps command during the Atlanta campaign, replacing Leonidas Polk upon his death.

- Lieutenant General Stephen Dill "S.D." Lee, 31, a native of South Carolina and a West Pointer. He was the youngest Confederate lieutenant general of the war. Lee served in the artillery and was captured at Vicksburg, then released. He was placed in command of cavalry in the Western Theater and assumed command of Hood's corps when Hood was elevated to army commander. Lee was a young but competent commander who nonetheless had been duly criticized for defeat at Tupelo (which was also one of Forrest's few defeats) and an ill-advised costly assault at Ezra Church (Atlanta). He was not related to Robert E. Lee.

The Army of Tennessee itself, although composed of seasoned veterans, was a hardluck bunch. Many of its officers and soldiers had been fighting since Shiloh in April 1862. As a colonel, Forrest served as the rearguard commander following Shiloh. Then and since, many officers and soldiers had been wounded, some several times. Some were missing an arm or a leg. Some had been captured and released from prison. Formed after the retreat from Perryville, Kentucky, in late 1862, the Army of Tennessee's first big battle came at Murfreesboro (Stones River), one of the bloodiest battles of the war, based on casualties as a percentage of men involved. After three days of a tactical draw, the Confederate commander, Braxton Bragg, withdrew from the field, giving the Yankees a major victory. Many of the men and junior officers in the Army of Tennessee sincerely believed that they had all but won the battles at Perryville and Stones River, only to have those victories denied them by the high command. Personal grudges and petty jealousies began to surface within the officer corps, straining the effectiveness of the chain of command. The army was outmaneuvered during the Tullahoma Campaign in the summer of 1863 and then forced to abandon Chattanooga, only to win its most glorious victory of the war at Chickamauga Creek, which then led to the siege of Chattanooga. At this point, U.S. Grant placed George Thomas in command of the Federal Army of the Cumberland. The subsequent victory at Lookout Mountain and the valiant charge up the heights of Missionary Ridge at Chattanooga sent the Army of Tennessee scrambling back into Georgia. Bragg relinquished command of the army to Joseph Johnston, a personal enemy of Jefferson Davis. In the early summer of 1864, Johnston, affectionately known to his men as Uncle Joe, and his army continuously fell back as

Sherman's three armies advanced through northern Georgia. Johnston was waiting for an opportunity to concentrate his forces and hit Sherman hard, but he ran out of time and territory. By the time the Yankees reached the Chattahoochee River in front of Atlanta, Davis had lost patience with Johnston and replaced him with Hood.

In hindsight, Hood's 1864 campaign was a dire, desperate move with little hope of success. Hood had lost 13,000 men in the attacks around Atlanta, leaving him with 30,000 effectives. Then he maneuvered and dawdled for eleven weeks before crossing the Tennessee River at Florence, north into Tennessee. Two weeks were wasted waiting at Tuscumbia for supplies to come by rail from Corinth and for Forrest to join him. "The campaign season was fast ebbing by the time Hood and his army got their act together," assessed historian Benjamin Franklin Cooling. Bad weather would play a major role in the upcoming operations. Hood also has been faulted for not paying attention to the details, all the minor decisions and careful preparations needed for any significant operation. Army of Tennessee historian McMurry cited Hood's "poor preparation, lack of attention to logistics, and poor reconnaissance." To feed his army, Hood would need 90,000 pounds of rations per day. Hood moved north into Tennessee with no more than half the supply wagons he needed for the army, according to historian and author James Lee McDonough. On the other hand, Hood's army was composed of 25 regiments of men from Tennessee, all enthused to be headed homeward on a mission to reclaim their homeland and rejoin their kinfolk.

The Federal authorities in Washington, D.C., and General Thomas in Nashville certainly took Hood's threat seriously. Federal troops from Missouri and Chattanooga were summoned to Nashville, as well as U.S. Colored Troops stationed at various garrisons. In the meantime, John Schofield and his small army were the only effectives between the Tennessee capital and Hood's army. In essence, Hood's northern advance to Nashville was a race with Schofield's small army.

Schofield, 33, was a native of New York and a graduate of West Point, where he had roomed with Hood as a cadet. Schofield was a career army man. He served for two years as an artillery officer and then procured teaching positions at West Point and

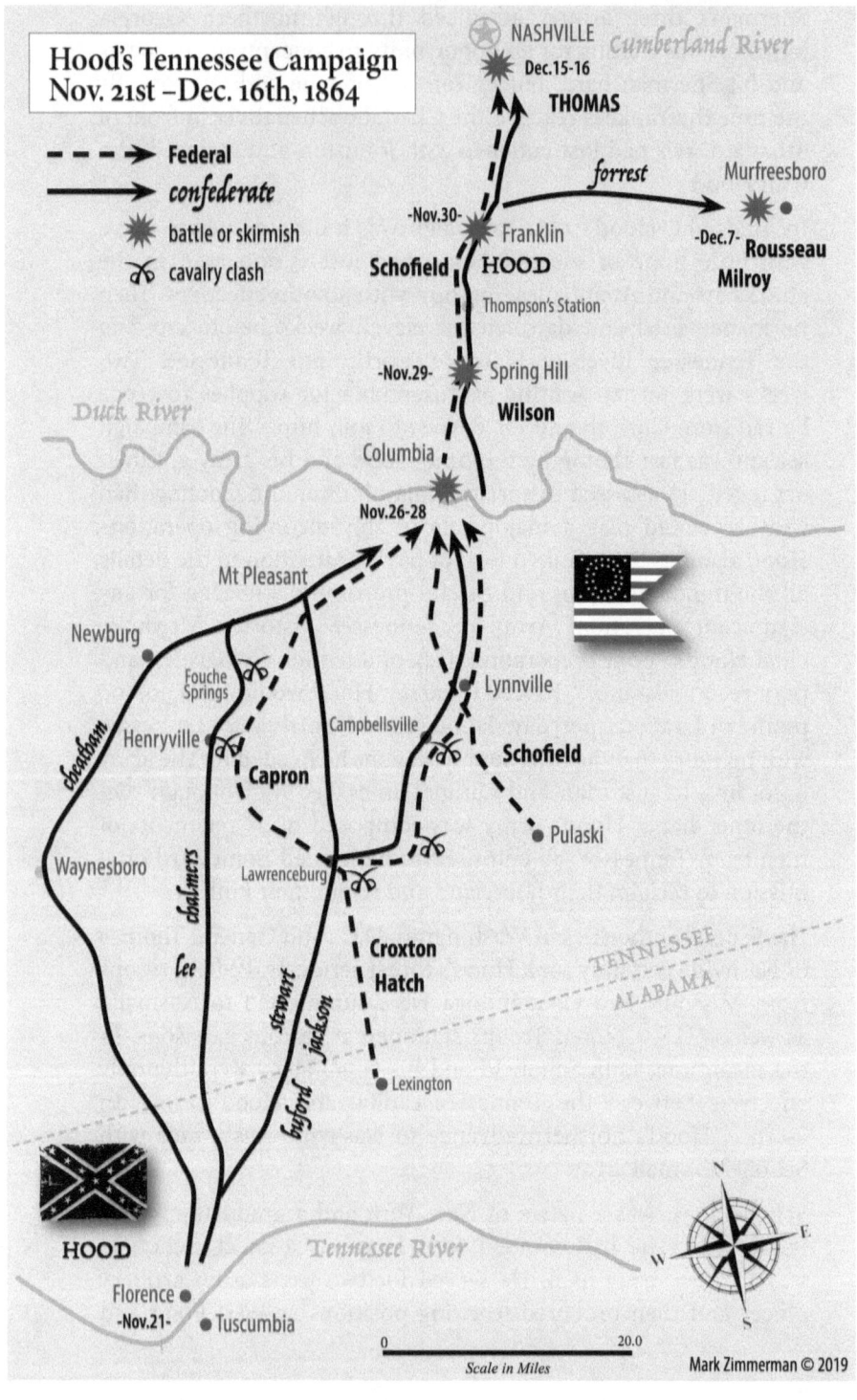

Washington University in St. Louis. With the outbreak of war, he was appointed chief of staff to Brig. Gen. Nathaniel Lyon. Schofield would ultimately be awarded the Medal of Honor for "conspicuous gallantry" at the battle of Wilson's Creek in August 1861. Promoted to major general of volunteers in November 1862, Schofield was made the commander of the Department of Missouri in 1863. He commanded the smallest of Sherman's armies in Georgia and was sent back to Tennessee when Sherman began his march to the sea.

Hood managed to outflank Schofield at Columbia, Tennessee, about 40 miles south of Nashville. Fighting broke out at Spring Hill but subsided at dusk. Hood had Schofield nearly surrounded and planned to finish him off the next day. During the night, however, in one of the strangest episodes of the war, Schofield's army marched north up the turnpike and through Spring Hill unmolested. When Hood learned the news on the morning of November 30th, he was livid. Most of the blame landed on the shoulders of Cheatham, one of Hood's three corps commanders. Hood moved his army northward on the pike to the outskirts of the small town of Franklin.

Schofield did not want to fight at Franklin, he only wanted to get his men to Nashville, the heaviest-fortified city in the nation (second only to Washington, D.C. itself). But he couldn't get across the Harpeth River fast enough. The turnpike bridge had been wrecked by the rising river and a second bridge damaged by fire. Only the railroad bridge remained sound. Three times, Schofield requested a pontoon bridge be brought down from Nashville, but the pontoon wagon train never arrived.

Schofield's troops, some armed with repeating rifles, stationed themselves behind formidable breastworks south of Franklin that connected the bends in the Harpeth River to the east and west of the village. Rejecting the strong advice of Forrest, who wanted to try a flanking attack, Hood launched his army on a frontal assault that resulted in one of the most magnificent displays of martial glory during the war. From Winstead Hill, the Army of Tennessee, flags fluttering, drums beating, bayonets gleaming, marched over more than two miles of pastureland exposed to the fire of the enemy. The forces involved, the casualties, and the distance covered exceeded those of Pickett's Charge at Gettysburg. Despite breaking through the Federal line at the Columbia Pike (site of the

Carter family farmhouse and cotton gin), the Confederates were driven back with horrendous casualties in six desperate hours of fighting, some hand-to-hand. In places, bodies fell stacked six deep along the breastworks.

"The Rebs came on to us in full force, and there ensued one of the hardest fought battles since this war commenced," wrote Lieutenant Alonzo Wolverton of the 20th Ohio Light Artillery. "The Rebs, determined to conquer or die, made thirteen desperate charges. Several times, they planted their colors within ten feet of our cannon, and our men would knock them down with their muskets or the artillerymen with their sponge staffs and handspikes. I never dreamed men would fight with such desperation. I never expected to come out alive."

Hood considered another attack on the morning of December 1st, but Schofield was gone. The Federals had crossed the river by planking the railroad bridge. The battle casualties, specifically the Confederate, were shocking—1,750 killed and 4,502 wounded or missing. Half of the regimental commanders (54) were killed or wounded. Fourteen of the 20 regimental commanders in Brigadier General Patrick Cleburne's division were lost. Six generals were killed (including Cleburne, known as the Stonewall of the West) or mortally wounded; one was captured; and five were wounded and out of action. A.P. Stewart lost five of his nine brigade commanders, and more than half of the colonels and majors commanding his 50 regiments. Six colonels, two lieutenant colonels, and three majors were killed. Fifteen colonels, nine lieutenant colonels, and six majors were wounded. Two colonels and two majors went missing. Following the battle, every structure in Franklin of any permanence, including all commercial buildings, churches, and many houses, 44 in all, were converted into makeshift hospitals.

After Franklin, brigades in the Army of Tennessee might be composed of up to ten small consolidated regiments and still be undersized, led by colonels or lesser rank, and rarely still led by their original namesake commanders.

Private Sam Watkins of the 1st Tennessee described the Confederate army following the debacle at Franklin: "Nearly all our captains and colonels were gone. Companies mingled with companies, regiments with regiments, brigades with brigades...I have never seen an army more confused and demoralized. The ground was frozen and rough, and our soldiers were poorly clad, while many, yes, very many, were entirely barefooted. The once proud Army of Tennessee had degenerated to a mob. We were pinched by hunger and cold. The rains, and sleet, and snow never ceased falling from the winter sky, while the winds pierced the old, ragged, grayback Rebel soldier to his very marrow. The clothing of many were hanging around them in shreds of rags and tatters, while an old sloughed hat covered their frozen ears. Some were on old, raw-boned horses, without saddles."

Hood could see no alternative other than following Schofield to Nashville and confronting the Federal army there. He sent Forrest with two-thirds of his cavalry command to attack the Fortress Rosecrans depot at Murfreesboro, which was manned by about 6,000 Federals. Hood then moved his men into weak defensive positions in a semi-circle south of Nashville designed to reach the Cumberland River to the east and west of the city. Hood did not have nearly enough men to cover that distance. The remaining division (Chalmers) of Forrest's command was used to protect Hood's flanks. Sending troopers under Forrest to skirmish with the Federals at Murfreesboro was Hood's "masterpiece of suicidal folly," according to historian Horn. Hood may have intended for Forrest's men to draw Thomas out of Nashville, but in any event, Forrest did keep thousands of Federals bottled up at Fortress Rosecrans.

By the time of the Nashville campaign, Forrest, 43, now a major general, was a living legend. He was less an expert at traditional cavalry and more a master at commanding mounted infantry (what used to be called dragoons), using terrain to his advantage, taking the offensive, moving aggressively, and employing psychological warfare against his opponents. He disdained the use of the saber, and armed his mobile troopers with sawed-off shotguns, revolvers, and short Enfield rifles. His artillerymen were highly proficient with the use of smoothbore canister but they also mastered the long-range effectiveness of the three-inch rifle (confiscated from the enemy). By the time of Hood's offensive, Forrest had gained a notorious reputation for his destruction of the Federal supply depot at Johnsonville, the alleged "massacre" of colored troops at Fort Pillow, and his raids against Federal garrisons at Memphis, Paducah, and Murfreesboro, just to name a few. News of a maneuver by Forrest's men was enough to send chills down the spines of Federal authorities and general officers. An imposing figure at six-foot-two, Forrest was a fierce warrior, earning the deep respect, if not the love, of his men. Born in Middle Tennessee, he grew up in the Memphis-Northern Mississippi area. His life was rough and tumble and his formal education lacking, but he yearned to become a man of means and social stature in his community. By the beginning of the war, he was one of the richest men in the South, having earned his wealth by speculating in land and prospering in one of the least respectable occupations—the marketing of slaves. He began the war as a private, then catapulted to the rank of colonel, spending much of his own money to outfit his cavalry units. In 1862, he declined to surrender his troops at beleaguered Fort Donelson and escaped through the winter weather to Nashville. A year later at Dover, he fought with his commander, demonstrating that he operated best when leading an independent command. During Hood's movement north toward Nashville, Forrest outfoxed the Federal cavalry at Columbia and Spring Hill. During the war, Forrest personally killed 30 men, including one of his own who had challenged him, and had 29 horses shot out from underneath him. For his prowess, he was known as the Wizard of the Saddle.

Forrest's command consisted of three divisions, led by Brigadier Generals James R. Chalmers, Abraham "Abe" Buford, and William Hicks "Red" Jackson. Forrest took Buford and Jackson with him to Murfreesboro and left Chalmers to assist Hood at Nashville. Hood also sent the infantry division of Major General William B. Bate to assist Forrest at Murfreesboro.

Bate's division consisted of Tyler's brigade of Tennesseans under Brigadier General Thomas Benton Smith, Finley's brigade of Floridians under Major Jacob A. Lash, and the brigade of Georgians under Brigadier General Henry R. Jackson. In addition, on the evening of December 6th, Forrest was reinforced by two small brigades of infantry—Sears' Mississippians and Palmer's Arkansans.

The Floridians had already seen a lot of fighting. General Jesse Finley had been badly wounded at Jonesboro, Georgia; his brigade was now commanded by Major Lash. All of Lash's regimental commanders were captains rather than colonels or majors. The men in the Florida brigade longed for home, cursed the bitter winter weather, and carried a grudge against General Bate for having stationed them in the precarious rifle pits at the base of Missionary Ridge the previous year. In the summer of 1864, the Florida brigade had earned the praise of General S.D. Lee for their fighting at Utoy Creek near Atlanta: "Soldiers who fight with the coolness and determination that these men did will always be victorious over any reasonable number," Lee said of them.

Forrest knew Murfreesboro well, having led a daring raid on the Federal garrison there in July 1862, earning him fame, respect, and promotion. Now, in early December 1864, he rode along the boundaries of the huge earthen Fortress Rosecrans, built in the spring of 1863 following the Federal victory at the Battle of Stones River. After examining the formidable works and its gun batteries, Forrest wisely determined it would be foolish to attack. He decided instead to draw the Federal troops out of the fort and into the open. That's exactly what happened on December 7th as a reconnaissance-in-force of 3,000 bluecoats was sent out under the leadership of Major General R.H. Milroy. The Federal leader and former Indiana lawyer was begging for a fight in order to redeem himself of charges made following the capture of his troops at Second Winchester in Virginia.

The so-called Battle of the Cedars did not go well for Forrest. The Confederate infantry, facing the enemy in its front, broke ranks and fled. These included the Floridians of Finley's brigade. Some of Jackson's brigade came up behind the Floridians, who were wearing blue Yankee jackets pilfered at Franklin, and may have opened fire on their comrades. Forrest said "the enemy moved boldly forward, driving in my pickets, when the infantry…from some cause which I cannot explain, made a shameful retreat, losing two pieces of artillery." Forrest and Bate tried unsuccessfully to rally the men. Reportedly Forrest ordered a color-bearer to halt, and when he didn't, Forrest drew his pistol and shot him. Forrest dismounted, took the colors, remounted, waved the flag, and finally rallied the men, according to Forrest's biographer, Brian Steel Wills. Milroy's men retreated back into the fort. Shortly thereafter, Bate's division of infantry was recalled to Nashville.

Bate was a complicated man, more of a politician than an officer, and a glory seeker who sought out the toughest assignments.

Private Philip Daingerfield Stephenson of the famed Fifth Washington Artillery of New Orleans stated: "Bate was a civilian, a politician pure and simple. In camp loose and careless; in battle short-sighted, ignorant and yet rash and ambitious. He was always thrusting himself forward, asking for the 'post of honor,' offering for specially difficult, delicate, dangerous work. Repeated blunders and failures had no effect on him apparently, either in making him wiser or more modest." He added, however, that Bate was "a fine person with manners and talents; he was genial, witty, and an effective stump speaker."

During the war, Bate was wounded three times and had six horses shot from under him. At Shiloh, Bate had been wounded in the leg. Doctors wanted to amputate, but Bate fended them off with a cocked pistol. From then on, he walked with a crutch and rode a white pony.

The Federal commander at Nashville, Major General George Henry Thomas, was a West Point graduate, veteran of the Mexican War and Indian fighting in Texas, but he was also a native Virginian. When the state had seceded and Thomas had remained loyal to the Union, most of his family had disowned him. During the war, Federal officials did not fully trust "the Virginian" and many superiors, including U.S. Grant, considered him to be too cautious and plodding. His nickname as an instructor at West Point had been "Slow Trot" and now at age 48 it was "Old Pap." His performance at Chickamauga in saving the routed Federal army had earned him the moniker of "The Rock."

"Few men have the qualities which deserve public confidence in greater measure than General Thomas," wrote Jacob Cox, a general officer who served under him. Cox described him as "a man of quiet, modest dignity who hated pretense and avoided notoriety." Thomas was "true to his superiors and kindly to his subordinates" and "oblivious of danger when duty required a risk to be taken."

Old Pap was known for his defensive tactics. But now he was anxious to show Grant and others that he could take the initiative in battle. He was looking for an opportune time to attack Hood outside Nashville, but he needed to wait for reinforcements. Then, the first two weeks of December, he needed to wait for the suddenly appearing icy winter weather to moderate.

From the nation's capital, Grant advised Thomas via telegraph: "Should you get him (Hood) to retreating, give him no peace." Grant's philosophy of war was simple: "Find out where your enemy is. Get at him as soon as you can. Strike him as hard as you can, and keep moving on." In the two weeks prior to the Battle of Nashville, dominated by severe winter weather, Grant prodded Thomas to get moving. Grant became so anxious, he sent General John Logan to replace Thomas and he threatened to go to Nashville himself. However, Thomas made his move before any of Grant's machinations could take effect.

On December 2nd, the veteran Federal XVI Corps of Major General Andrew Jackson Smith reached Nashville via riverboats from Missouri. Now reinforced with sufficient infantry, Thomas was confident of victory and optimistic that his well-drawn battle plan would work. But there was one factor he was less sure about. He needed to reinforce his cavalry.

Thomas thought Forrest had 12,000 men, twice Forrest's actual strength and six times the Confederate cavalry Hood had at Nashville. Thomas telegraphed Washington: "I have labored under many disadvantages since assuming the direction of affairs here, not the least of which was the reorganizing, remounting, and equipping of a cavalry force sufficient to contend with Forrest."

On December 2nd at Nashville, Federal cavalry commander Wilson was ordered across the Cumberland River to Edgefield to get his men rearmed, equipped and clothed, the horses shod, and to secure new mounts. At that time, 4,000 of his force of 14,000 troopers needed horses.

Throughout the war, horses were always in demand, for the cavalry and artillery, and to pull wagons. At the start of the war, the Northern states held approximately 3.4 million horses, while there were 1.7 million in the Confederate states. More horses were raised in Tennessee than any other state except Texas. The border states of Missouri and Kentucky had an additional 800,000 horses. There were 100,000 mules in the North, 800,000 in the seceding states, and 200,000 in Kentucky and Missouri. The disparity in the distribution of the mule population somewhat evened out the number of draft animals available for all purposes. An estimated 1.2 million to 1.5 million horses and mules perished during the war. Many were literally worked to death. The average lifespan of a horse during the war was seven months (normal lifespan was about 12 years).

Wilson spared no efforts in securing mounts and draft animals. Nashville's quartermaster was sent into Kentucky to buy mules. A traveling circus visiting Nashville lost its trick riding horses. Animals were confiscated from the stables of Military Governor Andrew Johnson, who reacted by calling young Wilson "a bumptious puppy." Nashville's streetcar line suspended operations because the mules were taken for the Federal army. Up north in Louisville, "horses were taken out of stables, street cars, wagons and buses, and … they were found in cellars, parlors, garrets, and all sorts of out-of-the-way places, where their owners had hidden them," according to Longacre. All these horses had to be shod and fed and perhaps curried. Saddles, blankets, and tack, bits and bridles, were required. Manure needed to be shoveled.

Nine days after moving across the river to Edgefield, Wilson issued Special Field Order No. 1 for movement back across the river at 8:30 am the following day. Two divisions would cross on the pontoon bridge; the remaining units would cross on the floored railroad bridge. The winter weather began to moderate on December 13th, and by noon on December 14th, a thick fog had burned off, the snow and ice were mostly gone, and the thawing ground became "a slushy, slick quagmire," according to historian Phillip R. Kimmerly, who conducted a detailed study of the weather and terrain at Nashville at the time.

The stage was set; the Confederates by now were eager to see action, any action at all. From their arrival at Nashville on December 2nd until the beginning of combat on December 15th, the Army of Tennessee suffered greatly from the harsh winter weather and the lack of provisions. Men dug shallow trenches and built small fires to avoid freezing to death at night. The daily ration was a quarter pound of graham bread, or two ears of corn, and a little beef. Some Confederates from the Deep South said it was the worst weather they had ever experienced. "The fact that they continued on without mass desertions or mutiny reflects well on their sense of duty," stated historian Massey.

## The Battle of Nashville Begins
*Thursday, December 15th, 1864*

On Thursday, December 15th, after several delays, most due to the weather, Federal commander George Thomas began the complicated process of moving thousands of infantrymen and cavalry out of the field fortifications around Nashville to attack Hood's men. Despite the dense fog, some confusion and glitches, Thomas' plan worked well that first day. The diversionary attack came first, against the Confederate right flank. The Federal forces, including U.S. Colored Troops, marched into a trap and were repulsed with heavy casualties. On the left flank, however, the grizzled veterans of General Andrew Jackson Smith's XVI Corps, along with Wilson's troopers, captured all five redoubts built by the Confederates along the Hillsboro Pike. The fighting between the flanks was sporadic. At dusk, Hood's army fell back two miles to the south into a compressed east-west line 2.5 miles long, anchored atop Compton's Hill to the west and Peach Orchard Hill to the east. During the maneuvering late on December 15th, Cheatham's Corps moved to the west flank, with Bate's Division atop Compton's Hill; A.P. Stewart's Corps crouched behind a stone farm wall in the center; and S.D. Lee's Corps fortified Peach Orchard Hill on the east flank.

Stewart's Corps had taken the brunt of the fighting the first day of battle. Colonel William D. Gale, assistant adjutant-general and aide to Stewart, observed the corps as it fell back into position. "I never witnessed such want of enthusiasm." The men seemed "utterly lethargic and without interest." Gale began to fear for the fight the next day and hoped that Hood would retreat.

On the second day of battle, Friday, December 16th, the Federals spent most of the day probing the hilly countryside for the new Confederate positions. General Thomas John Wood's IV Corps sparred with Stewart's Corps in the center while blueclad units tried unsuccessfully to turn S.D. Lee's eastern flank at Peach Orchard Hill.

The eastern or right flank of the Confederate line was anchored on top of Peach Orchard Hill adjacent to the macadamized Franklin Pike, the main thoroughfare south of Nashville. The Confederates on the hill under S.D. Lee repulsed attack after attack from the Federals and managed to hold firm control of the pike, the main escape route. The cavalry brigade of Colonel Jacob B. Biffle, consisting mostly of the 10th Tennessee, was stationed east of the pike and the railroad near the Travelers Rest estate of Judge John Overton.

Meanwhile, on the western flank, Schofield and Smith organized their Federal units to the northwest and northeast, respectively, of Compton's Hill. U.S. artillery units (each battery boasted six guns; Confederate batteries generally comprised four guns) began to bombard the hill as Wilson's troopers maneuvered around the hill and the smaller hills just south of it in an attempt to gain the Granny White Pike escape route.

Falling back from their positions near the Cumberland River on the far-left flank was James Chalmer's main force of Confederate cavalry under Colonel Edmund W. Rucker, consisting of the 1,500 troopers of the 12th Tennessee under Colonel U.M. Green; the 14th Tennessee under Colonel Rolla White; the 15th Tennessee under Colonel F.M. Stewart; and the 7th Tennessee under Colonel W.H. Taylor. On the evening of the first day of battle, the Federals had captured or destroyed many of Chalmers' wagons near the Belle Meade plantation.

Hood moved his headquarters from the Judge John Overton farmhouse (Travellers Rest) to the J.M. Lea farmhouse, and late on the first day of battle he watched the fighting from atop Compton's Hill. The hill was only about 160 feet high but quite steep, with two smaller hills to the south. Even farther south sprawled a range of much larger slopes known as the Overton Hills. What lingered at the back of the minds of most Southern soldiers on that left flank was their escape route running north-south just to the east of Compton's Hill known as Granny White Pike. The pike ran southward through the hills along Hollow Tree Gap and then intersected with the Franklin Pike. If Granny White Pike was blocked, the only other escape route was eastward along the northern fringe of the Overton Hills to the Franklin Pike.

An impressionable Federal soldier from Iowa claimed that Compton's Hill "in height and abruptness was almost a mountain," and Hood reassured his men that Compton's Hill was impregnable. Nonetheless, facing reality, Hood had already ordered his supply trains and wagons, excepting artillery, ordnance, and ambulances, to move five miles southeastward to Brentwood (a station on the railroad and the Franklin Pike).

Atop Compton's Hill, Hood could see Felix Compton's farmhouse to the northwest, near the now-defunct rebel redoubt No. 4. Hood swayed slightly in the saddle, his artificial right leg stuck tight into the stirrup, as he surveyed the battle scene with his field glass. He may have felt phantom pains in that absent leg, or perhaps joint pain in the shoulder of his useless left arm, supported in a sling. Having doffed his gloves, he may have noticed the scar on the back of his left hand and reminisced about his stint in the 2nd U.S. Cavalry at Fort Mason, Texas, seven years ago, before the war, when he served in a blue uniform under steadfast Major George Thomas. At one point, Hood was ordered to lead a small expedition down a Commanche trail in search of Indian warriors. After 12 days of riding, their uniforms gray from the dust, their throats parched from the lack of water, contact was made, and the horse soldiers found themselves in hand-to-hand combat with a band of Commanches. No quarter was asked nor given by either side. Armed with a dragoon saber, two Navy Colt revolvers, and a double-barreled shotgun, Hood dispatched several Indians before he was wounded. An arrow pierced both his left hand and the reins, pinning him to the saddle. He broke off the arrowhead, withdrew the shaft, wrapped a handkerchief around the wound, and continued his command. The troopers lost two killed and five wounded. The Indians suffered 19 killed and many others wounded. Three weeks later, the cavalry unit finally arrived back at Fort Mason after traveling a total of 500 miles. But that was all in the past.

Joining Hood on Compton's Hill was Ector's Brigade, an elite unit of Texans who had fought that first day isolated on the left flank, cut off near the Compton house. Although they had merely guarded wagons at the Battle of Franklin, they were used as a fire brigade at Nashville. Brigadier General Matthew Ector had been wounded at Atlanta and his leg amputated. His successor, Colonel Julius Andrews, was wounded at Bell's Mills in early December near Nashville. Now led by Colonel David Coleman, 40, of North Carolina, the understrength brigade consisted of six regiments of Texas cavalry, four dismounted, numbering about 100 men per regiment. Coleman attended the U.S. Naval Academy and served in the U.S. Navy during the siege of Vera Cruz, Mexico. He resigned from the navy in 1850, practiced law in Asheville, and served in the state senate. He favored secession and offered his services to the Confederate navy. No ship was available, so he organized Coleman's battalion, which became part of the 29th North Carolina, which now belonged to Ector's Brigade. As Ector's Brigade was passing him, Hood stopped them. "Texans, I want you to hold this hill regardless of what transpires around you." They replied, "We'll do it, General."

At 6:00 pm the first day of battle, Ector's Brigade tangled with Colonel John Mehringer's brigade of Schofield's XXIII Corps on Compton's Hill. In the pitch-black darkness, Mehringer mistook the rebels to be a much larger force and halted the attack. The first day's fighting on December 15th ended with winter's early nightfall saving Hood's army.

"The morning of December 16th found most of Hood's force exhausted, as would be expected of men who have fought all day and worked all night—hardly the ideal preparation for another day of fighting," noted historian Horn. At 8:00 am that morning, Ector's Brigade was moved from Compton's Hill to the smaller hills just to the south.

Due to the brief Mehringer-Ector firefight at dusk on December 15th, Federal corps commander Schofield became concerned about a counterattack by Hood on the 16th. Having attended West Point with Hood, he knew full well Hood's aggressive nature. Schofield was spooked, haunted by the recent events at Spring Hill and Franklin, and more concerned about defending than attacking. He positioned his divisions of Jacob Cox and Darius Couch to the west of Compton's Hill but he did not order an advance. Schofield reported, "I have not attempted to advance my main line today, and do not think that I am strong enough to do so." Meanwhile, Wilson's troopers skirmished with the Confederate troopers of

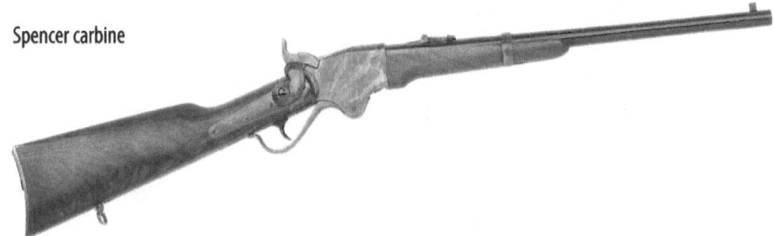

Spencer carbine

Rucker and, slowly but surely, the Federals worked themselves through the hilly, muddy terrain to the southwest of Compton's Hill in an attempt to block the Granny White Pike escape route.

Much like Forrest's men, Wilson's cavalry did not function in the traditional role of cavalry. He used his well-armed troopers as mobile mounted riflemen. Moving into position, the troopers would dismount and fight as infantry. In groups of four, three would dismount while the fourth man tended the horses, ready to move to another spot as needed. With their Spencer repeating rifles, the Yankees could apply almost overwhelming firepower.

"Cavalry is useless for defense," Wilson insisted. "Its only power is in a vigorous offensive...The true plan of action was a heavy dismounted skirmish line corresponding to the infantry line of battle, with a mounted force to charge the enemy's flanks and cut in upon his rear as opportunity offered."

Ironically, cavalry fighting as infantry, so effective during the decisive Battle of Nashville, would prove far less valuable during the subsequent fast-moving retreat. Many of the clashes would be on horseback, at close quarters, where use of the saber would be required.

Many of Wilson's troopers were armed with Spencer repeating rifles, the weapon you could "load once and fire all day," according to one Southern wag. Instead of loading paper cartridges down the barrel via the muzzle, like conventional longarms, the Spencer fired .52-caliber brass rimfire cartridges fed into the breech from a spring-loaded tubular magazine inserted through the buttstock. The cartridge contained the lead bullet, gunpowder, and primer all in one waterproof metal container. After shooting, the trooper cocked the hammer backward and then cranked the lever down and back into place to expell the spent cartridge and insert a new one. The trigger was then ready to be squeezed. The tubular magazines held seven cartridges; a Blakeslee cartridge box, carried via a sling, held ten preloaded magazines. A proficient trooper could fire 20 rounds a minute with a Spencer. The Spencer carbine, designed for cavalry, sported a 22-inch barrel versus the 30-inch barrel of the rifle and was two pounds lighter.

"There is no doubt," said Wilson, "that the Spencer carbine is the best firearm yet put into the hands of the soldier, both for economy of ammunition and maximum effect, physical and moral. Our best officers estimate one man armed with it is equivalent to three with any other arm. I have never seen anything else like the confidence inspired by it in the regiments or brigades which have it."

During the war, regiments firing the Spencer were often mistaken for brigades due to their superior firepower. The Spencer was faster to operate than the Sharp's or the Henry rifle and cheaper to manufacture (sold to the government for $40 each). The weapon was invented by Christopher Spencer, a Sharp's employee in Connecticut who won a military contract after President Lincoln test-fired the rifle and found it to his liking. The Spencer tended to shoot high, 12 to 18 inches at 50 yards, and was effective at only 500 yards or less due to its small powder charge. Troopers were not accurate at much more than 200 yards in combat, so this was a minor shortcoming.

Contrary to widespread belief, not all of Wilson's troopers at Nashville were armed with the Spencer repeating carbine. Many of the Federal cavalry were armed with Burnside, Maynard, and Sharp's carbines, evidenced by the recent findings of disparate spent ammunition upon the battlefield. Similar to the Spencer, these three types of lever-action carbines were breachloaders but with several crucial differences. The Burnside, Maynard, and Sharp's were single-shot rifles with no magazines. The metallic cartridges of the Burnside and Maynard did not incorporate primers and used percussion caps instead. The Sharp's used a paper cartridge. Although all of the carbines produced similar range and accuracy, the Spencer boasted a higher rate of fire. The Henry repeating rifle was not manufactured as a carbine and was not used by the Federal cavalry, although a few units of Federal infantry used a small number of Henry and Spencer rifles.

Matching the repeating rifle in firepower was the Federal cavalry commander who insisted his men be armed with it. At age 27,

Brevet Major General James Harrison "Harry" Wilson was one of the boy wonders of the war. A native of Shawneetown, Illinois, and an 1860 West Point graduate, Wilson stood erect at 5 foot, 10 inches tall. He was a "slight person of light complexion and with a rather pinched face," according to historian Jerry Keenan. He was also arrogant, highly ambitious, and outspoken, a "glory hunter of the first magnitude," according to biographer Longacre. Wilson first served as chief topographical engineer for the Federal Army of the Tennessee. He was an engineer during the Vicksburg campaign, and became good friends with U.S. Grant. He was the only officer from Grant's regular staff to be promoted to troop command.

In February 1864, Wilson became chief of the Cavalry Bureau in Washington, D.C. despite his having no special experience with horses. He proved to be an excellent organizer and administrator, and greatly helped the Federal cavalry grow and mature to the point where it matched or bested the Confederate cavalry, which had been the far superior force at the beginning of the war. In April 1864, Wilson was given command of Sheridan's 3rd Cavalry Division, his first independent field command. He made some costly mistakes but learned quickly. On September 30th, he reported to Sherman as chief of cavalry for the Military Division of the Mississippi.

Wilson was dedicated to succeeding at almost any cost, according to Longacre, who wrote, "He was merely a man who knew exactly what he wanted from life and believed he knew how to go about attaining it." Wilson did not accept blame for failure. "A determination to absolve himself of every vestige of responsibility for a flawed or questionable performance in command was one of Wilson's most enduring traits."

Wilson was eager to please his superiors. Historian Peter Cozzens noted, "Boasting came naturally to Wilson, whose undisputable military genius was liberally leavened with conceit, self-righteousness, and an unshakable certainty in the soundness of his plans."

He was a bachelor and a teetotaler. Longacre: "Since youth he had hated drinking with a passion and was convinced that under no circumstances — certainly none concerning an army in the field — did it serve a useful purpose."

Wilson was itching to face Forrest again. Two weeks prior, on the march north to Nashville, Forrest's three divisions clashed with Wilson's two at Hurt's Corner, six miles north of Duck River, and drove Wilson down Lewisburg Pike toward Franklin and away from the main action.

Longacre claims Wilson made a serious mistake at Columbia during Hood's advance: "In his eagerness to withdraw, Wilson made the most serious tactical error of his military career. Instead of falling back via routes which would enable him to fulfill his primary job—to serve as the eyes and ears of the army—he led his troopers up the Lewisburg Pike, in a direction that carried them far from Schofield's main force above Columbia."

At Spring Hill, Schofield wired Thomas that "I do not know where Forrest is...Wilson is entirely unable to cope with him."

At Nashville, Wilson was determined to rectify that failing. He was determined that his Federal cavalry would serve a major role in beating and destroying Hood's army.

Wilson's cavalry command, totaling more than 14,000 men, consisted of four divisions:

- 1st Division of Edward M. McCook, consisting of the 1st Brigade of Brig. Gen. John T. Croxton and his 1,300 troopers (the two other brigades were on a raid into Kentucky);
- 5th Division of Brig. Gen. Edward Hatch (4,621 men), consisting of the 1st Brigade of Col. Robert R. Stewart and the 2nd Brigade of Col. Datus E. Coon;
- 6th Division of Brig. Gen. Richard Johnson (4,034 men), consisting of the 1st Brigade of Col. Thomas J. Harrison and the 2nd Brigade of Col. James Biddle; and
- 7th Division of Brig. Gen. Joseph F. Knipe (3,867 men), consisting of the 1st Brigade of Bvt. Brig. Gen. John H. Hammond and the 2nd Brigade of Col. Gilbert Johnson.

During the Battle of Nashville and especially during the pursuit afterward, Wilson often broke up the divisions and used their brigades as independent units.

Edward Hatch, 31, a native of Maine, was one of Wilson's most trusted and effective subordinates. He was a handsome, talkative former sailor and lumberman. He moved to Iowa for the lumber business and became captain of the 2nd Iowa cavalry. He commanded a brigade at Corinth and participated in the famed 1863 Grierson Raid across Mississippi. In December 1863, Hatch tangled with Chalmers at Moscow, Tennessee, sustaining heavy losses. He was named brigadier in April 1864. Wounded during a West Tennessee raid, he supervised the cavalry depot at St. Louis before returning to the field. Just before Nashville, it was alleged that Hatch plundered the Oxford, Mississippi home of former U.S. Secretary of the Interior Jacob Thompson while Thompson's wife looked on. Hatch allegedly loaded an ambulance with pictures, china, glassware, and silverware.

Richard Johnson, 37, led the 6th Division. He was born in Smithland, Kentucky. His older brother served as a surgeon in the Confederate army. A member of the 1849 class at West Point, Johnson served on the frontier, and was appointed brigadier in October 1861. He surrendered to John Hunt Morgan at Gallatin, Tennessee in the summer of 1862. Exchanged in December, he commanded a division in the Army of the Cumberland at Stones River, Chickamauga, and Chattanooga. He was wounded in May 1864 at New Hope Church, Georgia. He was chief of cavalry in the Division of Mississippi before serving with Wilson.

Datus Coon, 34, of New York, had moved to Iowa and established several newspapers there. In July 1861, he enlisted in the 2nd Iowa Volunteer Cavalry and was commissioned a captain. He was with the 2nd Iowa Cavalry through all its campaigns from February 1862 to the end of the war, rising to the rank of colonel in command of the regiment.

John Hammond, 31, a native of New York City, had served as acting adjutant general and chief of staff for General Sherman. On Oct. 31st, 1864, he was made brevet brigadier general for services rendered.

Early on the morning of December 16th, the second day of the Battle of Nashville, Federal commander Thomas broke the stalemate on the Federal right flank by ordering Wilson forward at 9:30 am. Soon, Hammond held firm possession of the Granny White Pike at the gap south of Compton's Hill, with Hatch's division in cooperation. Croxton was close by with his brigade, in reserve. Wilson's force stretched 1.5 miles long, completely in the rear of Hood's line with 4,000 men (about the same number as Cheatham's entire corps). By noon, Wilson reported he had "a continuous line of skirmishers stretching from the right of Schofield's Corps across the Granny White Pike."

At noon, Coon's brigade was tangling with Chalmers' horsemen. Chalmers requested help, and Hood sent Ector's brigade from Compton's Hill. A prolonged skirmish ensued until around 3:00 pm, when Hatch called up the 1st Illinois Light Artillery, armed with two six-pounder mountain howitzers. After about 50 rounds had been fired, the Federal troopers charged and drove back the enemy force. The troopers consisted of the brigades of Coon, Stewart, and Hammond, totaling about 8,500 men.

Nature began to dictate its terms. Movement against the Confederate left flank was required soon, as the sun was setting quickly toward the southwestern horizon. The sun would set that day at 4:33 pm, with the moonrise at 9:09 pm. At that time of the year, there was only about nine hours of daylight to work with.

Horn summarized: "It was soon to develop that Wilson's cavalry, fighting dismounted and wielding the terrifically superior firepower given them by their Spencer repeaters, were able to supply Thomas with the difference between a stalemate and a smashing victory."

Wilson was well aware of what was necessary. Military genius Carl von Clausewitz had asserted in his *Principles of War*: "Only when we cut off the enemy's line of retreat are we assured of great success in victory." At noon, Wilson signaled to Thomas that he was ready to begin the main attack, but Schofield balked, asking for reinforcements. Wilson sent one staff officer after another to Thomas. Wilson said his men intercepted a courier's message sent by Hood to Chalmers exhorting, "For God's sake, drive the Yankee cavalry from our left and rear, or all is lost."

Besides Schofield, the other main force of Federal infantry facing Compton's Hill, from the northeast, was John McArthur's Division of A.J. Smith's Corps. A muddy cornfield of stubbed stalks stood between them and the foot of the hill. Rebel trenchworks wound down the hillside and joined the stone wall to McArthur's left, behind which crouched the enemy.

Brigadier General John McArthur, 38, was born in Erskine, Scotland, and emigrated to the U.S. at age 23 as a blacksmith and settled in Chicago, becoming a manager at the Chicago Iron Works. At the start of the Civil War, he became a captain of a militia company (Chicago Highland Guards), which was later incorporated into the 12th Illinois Regiment, which McArthur commanded as a colonel. He fought at Fort Donelson, where he was promoted to brigadier general, and was wounded in the foot at Shiloh, where his unit suffered horrendous casualties. He also fought at Corinth and led a division during the Vicksburg campaign.

McArthur was controversial, outspoken, and not very adept at politics. He had not been promoted above brigadier general since early 1862. He is not to be confused with the younger Arthur MacArthur, the Medal of Honor winner and the future father of World War Two General Douglas MacArthur. Arthur MacArthur had, in fact, been seriously wounded at Franklin and was resting at a hospital in Nashville.

Historian McDonough: "(John) McArthur was neither a career army man nor a West Pointer. But the Illinois general had a knack for command and an eye for terrain. He was a man accustomed to getting things done."

McArthur's Division consisted of three brigades of Midwesterners:

- 1st Brigade of Colonel William L. McMillen of Ohio;
- 2nd Brigade of Colonel Lucius F. Hubbard of Minnesota; and
- 3rd Brigade of Colonel William R. Marshall of Minnesota.

Most of the men in Thomas' army were from the states now called the Midwest — Ohio, Indiana, Illinois, Iowa, Wisconsin, Michigan, and Minnesota. There were also some units from Pennsylvania, Kentucky, and Missouri.

The men of Major General Andrew Jackson Smith's XVI Corps had come a long way to fight at Nashville. They had marched 800 miles in Missouri chasing Confederates, and reached St. Louis on November 24th, loading onto 59 river transports. They reached Nashville on December 2nd, General Smith receiving a welcoming bear hug from the usually taciturn Thomas. None had come farther than the four regiments from Minnesota, a state created only three years before the war. One regiment fought in the Dakota Indian War of 1862 before journeying south to fight the rebels. The Minnesotans of the 5th, 7th, 9th, and 10th regiments and the 2nd Battery of Light Artillery populated the three brigades of McArthur's Division.

It was determined that McArthur's men would take the direct route up the hill once the attack commenced, after they trudged through that muddy cornfield. As the stage was set at Compton's Hill, the blueclad riflemen knelt or squatted along the fringe of the cornfield, leaning on their weapons for support, smoking pipes or chewing tobacco, and double-checking their ammunition. Repeatedly, the percussion of artillery volleys rumbled among the hills, filling the ravines with smoke. Commanders waited impatiently for their couriers to return with word from headquarters. The command staff made for the ready; buglers bided time pursing their lips and spitting. Horses snorted, stomped, and scraped at the ground. Finally, anxious and worried, McArthur told one of his couriers to inform General Smith that he was ready to attack, and would attack, that damned hill unless he was directed to do otherwise.

On Compton's Hill, Private Stephenson of the Washington Artillery noted that he and his Confederate comrades began December 16th in a sullen and hungry mood. "We did want that miserable breakfast badly, for it is hard work fighting on an empty stomach, and we knew there would be no time for another meal that day. We clung to our frying pans, squatting around fires and bolting what food we could, but the shot and shell and bullets came booming and shrieking over our heads or skipping through our midst."

Positioned atop Compton's Hill, which flattened to a summit covering perhaps an acre, was Billy Bate's 1,500-man division of Cheatham's Corps. Tyler's Brigade was on the summit and down the western side of the hill; Finley's Brigade was positioned down the eastern side of the hill facing north; and Jackson's Brigade to its right and further down the hill. Situated along smaller hills running south from Compton's Hill were the brigades of Brown's Division. Even farther south were the brigades of Ector, Govan, and Reynolds, ever-increasingly refused to face southward against the encroaching advancement of Wilson's cavalry.

Pvt. Stephenson described what he saw south of Compton's Hill: "Here massed in great confusion so as to be hopelessly in each other's way, were most of the ordnance wagons and ambulances, as well as many of the quartermaster and commissary trains of our whole army! It... exposed the ordnance wagons to being blown up, and the ambulances to having their occupants torn to pieces."

The Confederate line east of Compton's Hill linked with the left flank of Stewart's Corps, Walthall's Division, consisting of the brigades of Sears, Cantey, and Quarles. Running north-south just to the east of Compton's Hill was the escape route of Granny White Pike. If the macadamized pike were blocked by Wilson's troopers, the only other escape for the Confederates would be a more southeasterly route over hilly terrain to the Franklin Pike. Direct passage to the south was blocked by the high chain of the Overton Hills. Five miles southeast of Compton's Hill was the Brentwood station on the railroad line and a small redoubt.

To make matters worse, the breastworks on the hill had been misconfigured. During the night, the Confederate engineers had mistakenly placed the breastworks on Compton's Hill too far up the slope, above the military crest. The enemy could advance up the hill without hardly being seen! As an officer on Compton's Hill would soon complain: "A six-foot-man could not be seen twenty yards from the front, thus rendering it possible to mass an attacking party within a few yards of the position and be perfectly sheltered from our fire."

Historian Horn elaborated: "By some engineering blunder, in the darkness and confusion of the preceding evening, the breastworks were placed so far back from the brow of the hill as to give the defending force a view and range on the front of not more than five to twenty yards. Thus, by this error, the steep face of the hill became rather more of an asset to the attackers than the defenders. Breastworks also were flimsy and there was no abatis (felled trees and branches) or other obstructions."

In addition, tactical maneuvering by the high command siphoned units off the hill, moving them farther south or even to the right flank, leaving Compton's Hill dangerously undermanned. About noon, Hood needlessly ordered two of Cheatham's brigades to reinforce Peach Orchard Hill, leaving only Govan's Arkansas brigade to patrol an area south of Compton's Hill originally covered by an entire, albeit depleted, division. At 3:30 pm, the federal cavalry began driving a wedge between Govan and Coleman (Ector's Brigade). Hood ordered Brig. Gen. Daniel H. Reynolds' brigade out of A.P. Stewart's line to reinforce Coleman. Reynolds ended up about 300 yards east of Granny White Pike along a ridgeline facing north.

Private Sam Watkins of the 1st Tennessee noted, "We are continually moving to our left. We would build little temporary breastworks, then we would be moved to another place. Our lines kept widening out, and stretching farther and farther apart, until it was not more than a skeleton of a skirmish line from one end to the other."

The disparity in unit strength between the Federals and the Army of Tennessee must be noted. In some cases, a single Northern regiment mustered more men than an entire Southern brigade. The total effectives present in the nine regiments of Maney's brigade was 654, an average of only 73 men per regiment versus the normal number of 1,000. The 49th Tennessee Regiment of Quarles' Brigade in Walthall's Division was down to 17 men after Franklin. All in all, Hood was outmanned nearly 3-to-1 at Nashville.

At noon, it began to rain and Federal artillery batteries totaling nearly 100 guns to the north and west opened up on Compton's Hill. One Confederate private said, "The Yankee bullets and shells were coming from all directions, passing one another in the air."

Compton's Hill was subjected to a heavy artillery crossfire from three directions. Batteries located to the side of the hill on Schofield's right took Bate's first brigade in reverse, and the guns in the rear of the Bradford house threw shells directly into the back of Bate's left brigade. The barrage was intensified by the point-blank fire of Couch's artillery. One battery of the Indiana Light Artillery fired 560 rounds into the hill in three hours, pounding the works on the left of the angle for 50 to 60 yards. Another Indiana battery launched 923 shells, case, and solid shot against the hill.

"At one time the fire of at least twenty guns was concentrated on our position...Shells and canister were everywhere, the air was full of them...Five or six passed through a tree behind which I was sitting, every time approaching nearer my person, until at last I moved out," reported James Cooper, aide to General Thomas Benton Smith, the commander of Tyler's Brigade. The Yankee fire was so heavy that the Confederate cannon atop the hill had to be reloaded by drawing them aside with the stout rope or prolonge, to the protection of the parapet. "Three or four batteries were playing upon the few acres about the top of the hill, and if a man raised his head over the slight works he was very apt to lose it."

One Confederate infantryman said, "The cannon balls flew thick and fast, bursting all around, tearing up the ground, and cutting trees to pieces." Another testified: "To lie under a destructive artillery fire produces a feeling of dread that...is more demoralizing than to be actively engaged in battle; but we had to endure it for six long hours without even a skirmish to divert our attention. Our artillery replied only occasionally."

Field artillery during the Civil War was direct-fire — gunners could shoot only at targets they could see; there were no aerial spotters or forward observers to call in the coordinates as in 20th Century conflicts.

In response, Cheatham's Corps manned 34 field artillery pieces. Bate had a battery of howitzers under Captain Rene T. Beauregard (son of the famous general) on the eastern slope; Major Trueheart's two 12-pounder guns were added later. Lieutenant William Turner somehow got his Mississippi battery, attached to Brown's Division, onto one of the hills south of Compton's. At mid-afternoon Bate placed Beauregard's guns on Finley's line so as to cover the front of the Floridians and the left of Walthall's division.

During the exchange of artillery, the Federal gunners faced a danger of their own. Bate's sharpshooters up on the hill, commanded by Lt. A.B. Schell and armed with Whitworth rifles, harassed them and made life miserable for them with their accurate shots.

Private Stephenson noted: "Throughout the whole of that terrible day a spell seemed to be on the whole army, officers and men, a palsy not of fear, but of despair. All movements were mechanical, in a spirit of passive indifference like condemned criminals waiting the hour of execution."

At 3:00 pm, as the shadows lengthened, McArthur sent word to his commander, A.J. Smith, that his division could carry Compton's Hill by direct assault. He was told to wait, that Schofield was not ready. McArthur later claimed he did not receive the directive. He began to prepare for the assault.

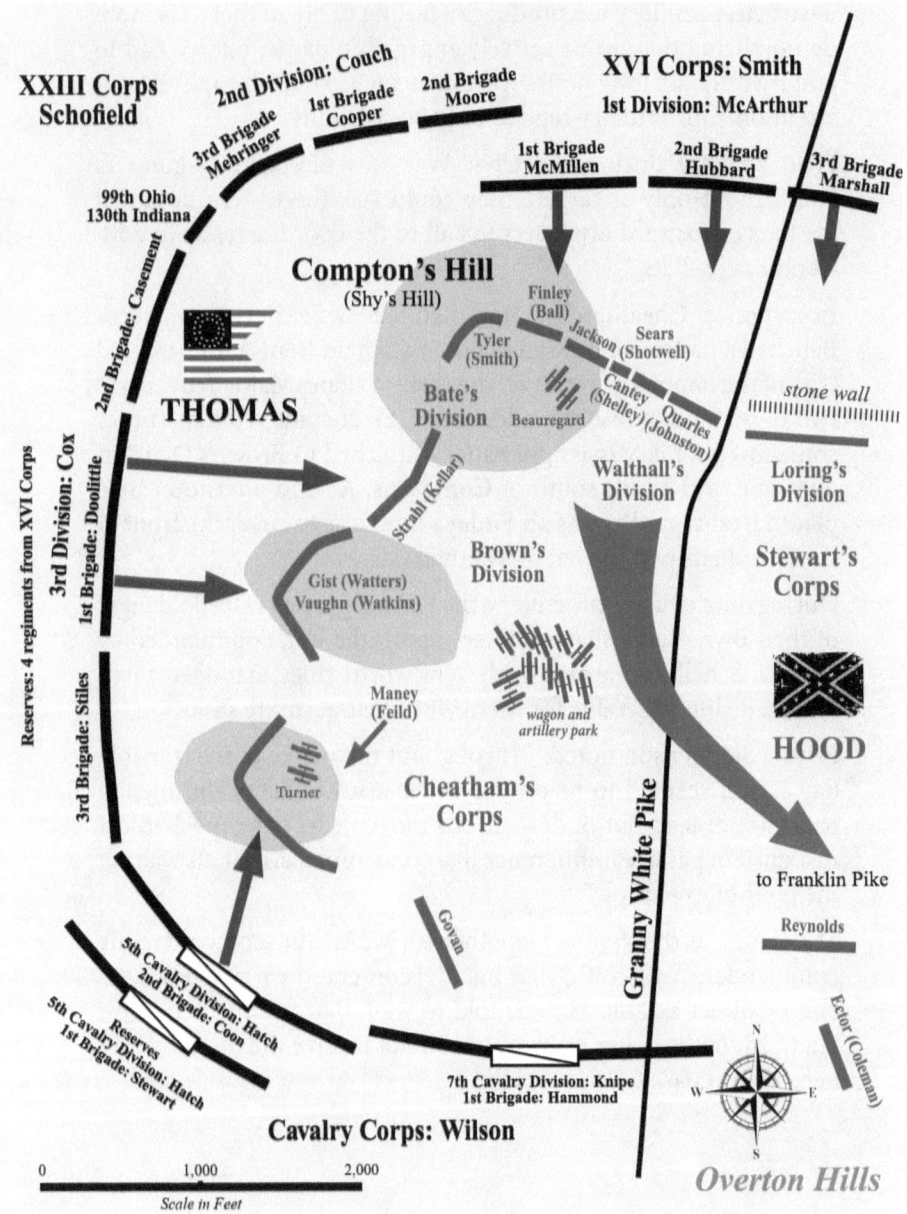

## Compton's Hill
*Friday, December 16th, 1864*

Finally losing his patience, John McArthur could wait no longer. The Scotsman rode up to McMillen, his first-brigade leader, and told him, "Take that hill!" McMillen pulled back his skirmishers, withdrew his regiments from building defensive rifle pits, and marched 500 yards to his right, halting in front of Couch's Division, where he formed his brigade into two lines of battle. At the last moment, Cogswell's Independent Battery of Light Artillery, McMillen's lone supporting battery, was discovered to be nearly out of ammunition. An attempt to replenish their supply from ordnance wagons in the rear resulted in mixed communications, and the pause became a delay. The lieutenant commanding Cogswell's battery hastily gathered ammunition from a nearby battery, enough to fire ten shells from each gun. Within a few minutes, McMillen was alerted and he ordered Cogswell's battery to open fire.

Following the 3:00 pm attacks on Peach Orchard Hill, Pap Thomas rode to the west flank and consulted with Schofield and Wilson. With a low growl, he advised them, "The battle must be fought, if men are killed." Then they noticed McArthur moving against the hill and Thomas told Schofield, "General (A.J.) Smith is attacking without waiting for you; please advance your entire line." By the time Wilson galloped away and reached his men, the rout was on.

Suddenly, bugles sounded and three batteries totaling 18 guns fired all at once, filling the hills with percussive waves and deep subsonic echoes. Birds dropped in mid-air, as if killed by the change in air pressure, a Federal chaplain noted. Row upon row of blueclad soldiers marched forward as rapidly as the mud would allow, flags snapping and bayonets jabbing the cold air, canteens clanking. In the memories of Federal veterans, the assault on Compton's Hill became one of the war's most glorious moments.

McArthur's men charged 400 yards over the open and muddy fields under heavy fire. McMillen's 1st Brigade, with the 10th Minnesota on the left of his first line, went first. Near the crest, the 10th Minnesota received heavy fire from their left flank, suffering 70 casualties, its commander, Lt. Col. Samuel P. Jennison, mortally wounded. Hard charging were McMillen's other regiments—the 114th Illinois and the 93rd Indiana, backed by the 72nd and 95th Ohio.

When McMillen was halfway up the hill, Hubbard's 2nd Brigade followed with the 5th and 9th Minnesota forming the front line. Marshall and his third brigade took up the charge when he saw Hubbard moving on his right. The 7th Minnesota was on the left of Marshall's first line. And about this time Cox's division in Schofield's Corps moved forward. About 3:00 pm, Hatch moved his Federal cavalry division along a ridge that ran south along the Confederate left and brought up Battery I, First Illinois Light Artillery, to open enfilading fire.

"It was more like a scene in a spectacular drama than a real incident in war," recalled Colonel Henry Stone of Thomas' staff. "The hillside in front, still green, dotted with boys in blue swarming up the slope, the dark background of high hills beyond, the lowering clouds, the waving flags, the smoke rising slowly through the leafless treetops and drifting across the valleys, the wonderful outburst of musketry, the ecstatic cheers, the multitude racing for life down into the valley below—so exciting was it all that the lookers-on instinctively clapped their hands as at a brilliant and successful transformative scene; as indeed it was."

General A.J. Smith said: "The enemy opened with a fierce storm of shell, canister, and musketry, sadly decimating the ranks of many regiments, but nothing save annihilation could stop the onward progress of that line." The blizzard of canister came from the Point Coupee (Bouanchaud) Louisiana battery behind the farm stone wall.

Of the 315 men in Hubbard's 2nd Brigade killed or wounded during the Battle of Nashville (out of a total of 1,421 engaged), most fell at Compton's Hill. The 5th Minnesota took 100 casualties, with three color bearers shot down and killed. Hubbard recalled what happened when the bugles blared forth at 3:00 pm. "I was ordered to storm the works in my front and carry the position at all hazards. The line of my advance lay across an open field about 400 yards wide in which every man was exposed to the deadly range of the enemy guns. I ordered the charge and moved my men forward. They obeyed promptly, advancing steadily and with the cheer presented a definite front to the storm of great canister and musketry that swept across that field. For a moment I felt a doubt, the fire was fearful and my lines seem to be melting away. The batteries of the enemy threw their grape with deadly aim. Great gaps were made in the ranks, my regiments were growing small, but they still advanced. We were suffering fearfully, but we were gaining the works. My horse was shot, gave a plunge, and fell dead. I mounted another and he too was shot. My staff was falling about me, I was struck by a minié ball on the neck and knocked headlong to the ground but was soon on my feet again. One hundred men of my old regiment lay upon the field dead or disabled, three hundred had fallen in the brigade. It was terrible, but the glorious fellows did not falter. Still on, and the works are gained, the parapets scaled, and the enemy drop their arms in surrender or fly in confusion to the rear. The victory is won."

Lt. Col. William B. Gere of the 5th Minnesota, who was awarded the Congressional Medal of Honor for capturing a battle flag: "…about 3:30 we assaulted their position; a fearful charge, hundreds fell, but we captured the works with prisoners by the thousands… The fighting was the heaviest in our front—it was indeed a desperate thing to go through, that storm of grape, canister, and musket balls. We who got through wonder how we escaped! I was lucky enough to get the battle flag of the Fourth Mississippi Regiment in the charge."

Marshall, who attacked just east of Granny White Pike: "The Federals were upon the Confederates so quickly and in such numbers that the Southerners immediately east of the pike were quickly overcome."

On top of Compton's Hill, Brig. Gen. Thomas Benton Smith of the 20th Tennessee commanded Tyler's Brigade. A graduate of Nashville Military Institute, he was the army's youngest brigadier general at 26 years old. At Stone's River, two years prior, the 20th Tennessee suffered 55 percent casualties, including five of their six color bearers, one of whom was Smith's only brother. T.B. Smith was colonel of the regiment at the time. He was severely wounded, a bullet passing across his chest and through the left arm. He was again wounded at Chickamauga.

At Nashville, Colonel William M. Shy, 24, led the 20th Tennessee. Born in Kentucky, he had lived with his family in Franklin since 1848. Unmarried, he was modest, calm, and collected. His mother sided with the South, his father with the North. Shy enlisted as a private in Co. H of the 20th Tennessee and was named to the color guard. He was promoted to major in 1863 and soon afterwards to lieutenant colonel. After Brig. Gen. Robert C. Tyler lost a leg at Missionary Ridge, T.B. Smith succeeded to brigade command; Shy then assumed command of the 20th Tennessee. In September 1864 at Atlanta, Shy was promoted to colonel and Smith to brigadier general.

Once the Federals had crossed the cornfield, scaled the slope, and crested the brow of Compton's Hill, the fighting became hand-to-hand, with the jabbing of bayonets and the pounding of rifle butts. Officers slashed with their sabers and discharged their revolvers at point-blank range. Many thrashed and wrestled on the soggy ground amidst the debris, yelling and screaming.

Colonel Shy defended the hill with his life, fighting to the end and falling from a close-range bullet wound to the forehead. The trajectory of the shot was downward. Half of his men were killed or disabled; 65 managed to escape. So heroic was the colonel's stand that Compton's Hill was renamed in his honor and has been known ever since as Shy's Hill.

James L. Cooper, an aide to T.B. Smith, said: "What had been feared all day occurred. A large force of the enemy massed under the crest of the hill, and dashed over the flimsy works before some of the men had time to fire a single shot. More than half the brigade was killed, wounded, or captured in a hand-to-hand struggle."

Fighting furiously, General Smith realized that his unit was surrounded by blue-jacketed Yankees. He waved his white handkerchief over his head and ordered his men to cease firing, according to Dr. Deering Roberts, the surgeon in Bate's division. A private in the 20th Tennessee said that General Bate, riding his white pony, told General Smith to get his men out, if possible, then

"Battle of Nashville" detail, mural (1906) by Howard Pyle, now exhibited at the Minnesota State Capitol, of Federal troops from Minnesota storming the Confederate lines east of Compton's Hill.

rode off and narrowly escaped capture himself. The private said, "The only reason that General Tom Smith and the few near him were not also killed was that the Yankees had passed around our left and were closing in on us from both front and rear, hence they could not fire on us without killing their own men."

Federal brigade commander Col. McMillen was a doctor from Columbus, Ohio. He had served as a surgeon for the British in the Crimean War. He was known as a drinker; he had been drunk at the Federal defeat at Brice's Crossroads in Mississippi. In high temper, McMillen, whose brigade had suffered heavily in the attack, reportedly berated Smith and then attacked the Confederate general, now a disarmed prisoner, with Smith's own sword, beating him on the head with the flat of the blade. Smith's head injuries were so severe (the brain was visible), it was feared he would not survive. The Confederate general somehow did survive his wounds, and was eventually imprisoned at Johnson's Island in Ohio and later at Fort Warren in Massachusetts.

Lt. Thomas Shaw of the Cumberland Rifles, 2nd Tennessee, stood his ground on Compton's Hill until he was slammed down and pinned to the ground by a Yankee bayonet. He was taken prisoner and soon died of his wound at a Nashville hospital.

Then perhaps the inevitable happened. At some point, the Confederate line atop Compton's Hill broke, causing a chain reaction of men fleeing toward the rear, without orders, many throwing down their weapons, trying to escape the onslaught. Nobody knows for certain, but many probably decided, after enduring hours of an artillery barrage, that they had come a long way during the war and did not want to die defending an insignificant hilltop against overwhelming odds in the service of a desperate cause.

Surrender was an option, of course, but in the mind of the ordinary infantryman spending the winter in a Northern prisoner-of-war camp was tantamount to a death sentence. Lt. Henry W. Reddick of Finley's Brigade: "I thought several times that I would have to fall out for I was completely broken down, but when I would think of being captured I would come to a new life."

Tales of torture and starvation and deprivations beyond belief (real or exaggerated) at prison camps had filtered their way back to the front lines throughout the war. Andersonville was the archetype of misery, but many other camps were just as bad. At Camp Douglas in Chicago, one out of every ten men died during the winter months, with a death toll of 4,000 to 6,000. Upon inspecting the camp, the U.S. Sanitary Commission reported that "…the amount of standing water, of unpoliced grounds, of foul sinks, of general disorder, of soil reeking with miasmic accretions, of rotten bones and emptying of camp kettles…was enough to drive a sanitarian mad." The barracks were so filthy and infested the commission claimed that "nothing but fire can cleanse them." More Confederate soldiers were buried in Chicago than any other place north of the Mason-Dixon line. Confederate Mound at Oak Woods Cemetery is the largest known mass grave in the Western Hemisphere. Perhaps even more notorious was the POW camp at Alton, just north of St. Louis, where more than 300 prisoners died during the winter of 1862 from a smallpox epidemic. Confederate officers were treated much better than non-coms, usually confined at camps such as Johnson's Island in Lake Erie or Camp Chase in Columbus, Ohio. Despite the horror stories, many men were exchanged and came home to fight again, especially during the first years of the war. On both sides, 56,000 men died in prison camps over the course of the war, accounting for roughly 10 percent of the war's death toll and exceeding American combat losses in World War I.

J.P. Cannon of the 27th Alabama said, "Lieutenant Colonel Weeden, of our regiment, ordered all who could to save themselves. It looked like escape was impossible, and most of the regiment stacked arms and surrendered; but dreading a Northern prison so much, I determined to make the attempt, and struck out obliquely to the rear as fast as I could run."

Some units were ordered to retreat. Lt. Col. James R. Binford of the 15th Mississippi, Stewart's Corps: "Realizing now I would not be reinforced and if I remained there much longer, myself and command would either be captured or killed, I gave the command to retreat by the right flank moving towards the Franklin Pike." They didn't move fast enough — 52 of Binford's men were subsequently captured. One private said retreating was like a living nightmare in which you just couldn't run fast enough.

In his memoirs, Pvt. Sam Watkins described the scene: "Finley's Florida brigade had broken before a mere skirmish line, and soon the whole army had caught the infection, had broken, and were running in every direction. Such a scene I never saw. The army was panic-stricken. The woods everywhere were full of running soldiers. Wagon trains, cannon, artillery, cavalry and infantry were all blended in inextricable confusion." He lamented, "The once proud Army of Tennessee had degenerated to a mob."

General Bate reflected later, "Whether the yielding of gallant and well-tried troops to such pressure is reprehensible or not, is for a brave and generous country to decide."

There was some dark humor. A young staff officer who had been in the rear rode among the fleeing men, shouting, "Stop! Stop! There is no danger there." A veteran soldier looked up at him and said, "You go to hell—I was there."

Observing the rout, U.S. General Jacob D. Cox of Schofield's command noted that "the enemy was manifestly disconcerted."

The Confederate commander of Bate's 3rd Brigade, Brig. Gen. Henry R. Jackson, 44, was a Georgia native, graduate of Yale, lawyer and state judge, veteran of the Mexican War, and Minister to Austria during the 1850s. He was also a poet and prominent public speaker. He saw action in West Virginia and led a brigade during the Atlanta campaign. Jackson tried to escape the rout at Compton's Hill and got to Granny White Pike, but he stumbled when the stone wall he was climbing upon collapsed beneath him. His boots got stuck in the mud. He was taking them off when he heard, "Surrender, damn you!" He and his aide saw four Yankees on a fence row aiming at them. The aide surrendered them. As Jackson put his boots back on, the aide turned down the general's collar to hide his insignia. Jackson turned the collar back up. "You are a general!" one of the Yankees said. "That is my rank," Jackson replied. The Yankee said, "Captured a general, by God! I will carry you to Nashville myself."

Federal Colonel Charles C. Doolittle was authorized by his commander to advance as soon as the crown of Compton's Hill was occupied by McArthur's troops. About 4:30 pm, Doolittle ordered his brigade eastward against the hill immediately south of Compton's Hill, topped by breastworks. Leading the way was the 12th Kentucky, followed by the 8th Tennessee and the 100th Ohio. The Confederate cannoneers abandoned their guns with the charges still in the barrels. Eight cannon and 300 prisoners were taken. Resistance was light. Doolittle suffered nine wounded, none killed.

Despite the debacle, some Confederates fought valiantly. Major Trueheart's artillerymen coolly and deliberately directed fire into the advancing Federals until they were surrounded and captured. They had no time to spike or disable their guns. They then witnessed a great indignity as their guns were turned on their fleeing countrymen. The Washington Artillery apparently had more time. Although they tried to save their guns, the commander was told by his superior to save his men, limbers and teams, and abandon the guns. The guns were spiked, dismounted, and the wheels destroyed. Then they ran for their lives with the enemy firing at them from three sides.

On the other hand, some Confederates put up no fight at all.

Colonel Andrew J. Kellar, commanding Strahl's Brigade of Tennesseans, stated that his men scattered from the vicinity of Compton's Hill without making a stand. Two days later he wrote a formal apology. "It was not by fighting, nor the force of arms, nor even numbers, which drove us from the field."

Confederate commander Hood was at his headquarters in the Judge J.M. Lea house, watching the debacle, when he heard the news. He claimed complete surprise; he was making plans for the next day's offensive actions. Hood recalled, "Our line, thus pierced, gave way, and soon thereafter it broke at all points, and I beheld for the first and only time a Confederate army abandon the field in confusion." Hood dispatched a courier to tell cavalry commander Chalmers to hold Granny White Pike "at all hazards."

Col. Gale, aide to A.P. Stewart: "On I struggled until I, too, became exhausted and unable to move. By this time the enemy had gotten to the foot of the hill and were firing at us freely…I twisted my hand in my horse's mane and was borne to the top of the hill by the noble animal, more dead than alive. I was safe, though, and so were my men. We descended the southern slope and entered the deep valley, whose shadows were darkened by the approaching night. The woods were filled with our retreating men."

R.B. Meadows of the 35th Alabama said, "When I reached the hill after a half mile run, I was almost ready to fall, but I kept going. I could hardly creep, but every once in a while a bullet singing around me would give me renewed courage to go a little faster."

The 5th Washington Artillery proceeded south down Granny White Pike to ascend the ridge and into a defile when the column halted, and "then a wild stampede came back upon us, everything that was in front of us, wagons, ambulances, horsemen, footmen, each striving to pass the other, jamming together, choking the road, yet ever struggling frantically back as if the legions of hell had met them!" What the Confederate fugitives had run into on the pike was Wilson's troopers (the 7th Division of Brig. Gen. Joseph F. Knipe). All movement was then channeled to the east.

Private Stephenson recalled: "After the most strenuous exertions of both horses and men, we succeeded in dragging our heavy Napoleons through gullies, marshes and thickets to a point about half way to the Franklin Pike. The bewildering disorder, the

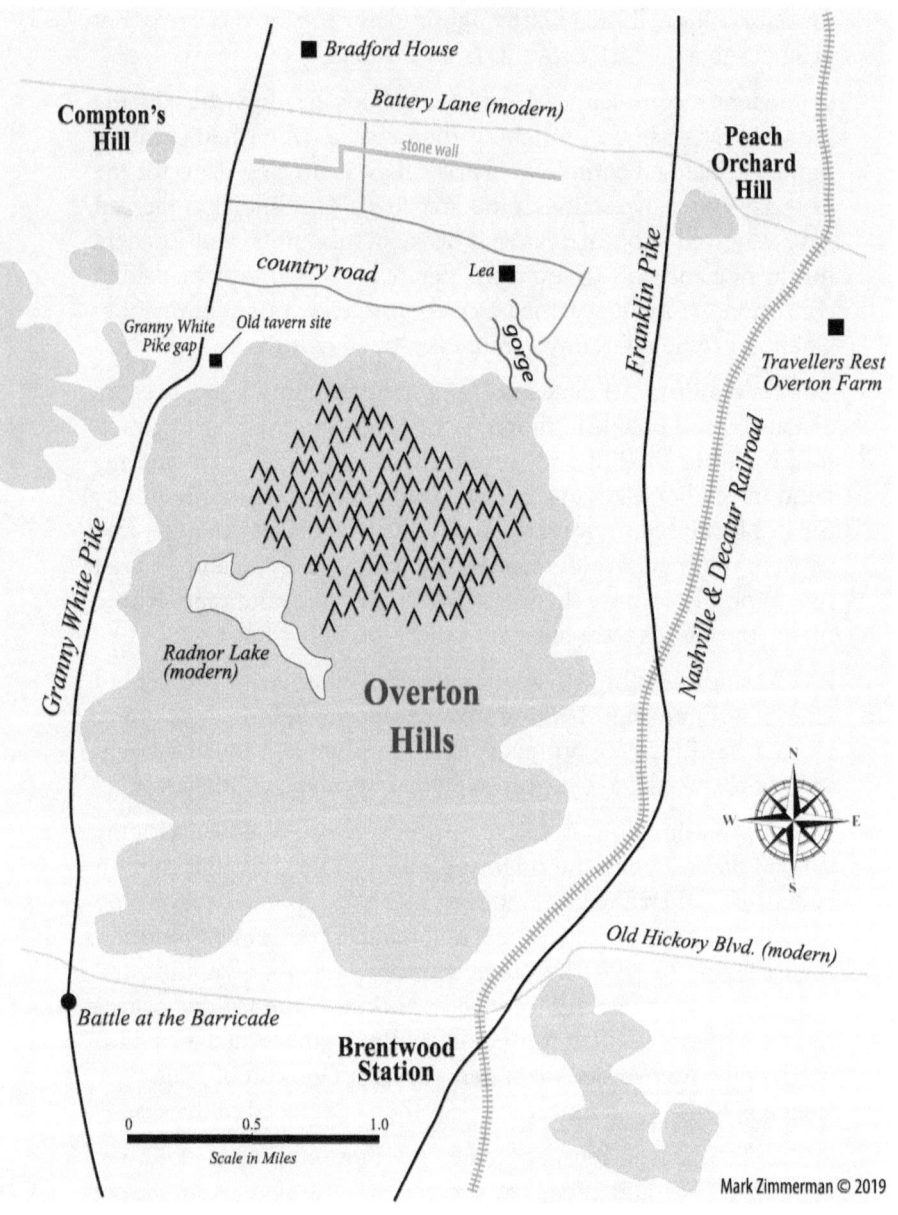

constant upsetting of wagons and consequent choking up of the way, the deafening uproar, the wild countenances of men, the almost hopeless chance of escape, these things robbed the stoutest of hearts of fortitude.

"In one awful hell of confusion lay the wagon train jammed against trees, tangled in the underbrush, sunk in the mire, and upside down. Mixed up with them, or flying from them, were our panic stricken men…We faced defeat, disgrace, death, or a loathsome prison… All things were at once abandoned. The wildest confusion and terror took hold of men. Each individual looked to himself, subordination and obedience were thrown to the winds… The blanched faces of the teamsters as they fumbled with nerveless hands to cut the traces of their maddened animals, the plunging steeds of the officers, the incoherent cries of our panic stricken infantry as they rushed among us, the piteous entreaties of our wounded not to leave them, the vast heap and ruined debris of the army machinery all around us, the fearful storm above and about us. Ah, that was a picture!"

Stephenson was lucky. Amid all the panic and confusion, he managed to capture a mule and ride it tens of miles southward to safety over the next few days.

Fortunately for the routed Army of Tennessee, darkness was rapidly falling, as was the cold hard rain, and Wilson's troopers could not immediately pursue in any great number. They had been fighting dismounted, and it took time to get mounted and organized for a pursuit. And the sudden rout of the rebels had taken the cavalry commander by surprise. Wilson later explained: "Rapid movements across rough country and plowed fields in the dark were impossible. Our troopers could hardly see their horses' ears."

Just south of Compton's Hill, Govan's Brigade, composed of the 600 men of 10 undersized Arkansas regiments, temporarily stopped the tide of Wilson's troopers from gaining the Granny White Pike (the three other brigades in the division had been moved needlessly to the right flank and eventually ended up in Brentwood about 3:30 pm). About noon, Govan had been wounded in the throat and the next ranking colonel was also hit. Maney's Brigade filled the gap and temporarily restored order. Govan had been an outstanding commander in Patrick Cleburne's shock troops, performing brilliantly at Chickamauga and Missionary Ridge (covering the retreat there). During Jonesboro, Georgia, Govan experienced his first real defeat when his command was overrun. He and several hundred of his men were captured, and then exchanged in September. At Franklin, Govan expressed his skepticism of Hood's plans to attack frontally, to which Cleburne famously replied, "Well, Govan, if we are to die, let us die like men."

Colonel Luke W. Finlay of the 4th Tennessee of Brown's Division watched Compton's Hill from the south and saw the U.S. flag on top. Concluding that the situation was hopeless, he told his men: "Make for a gap in the hill on the opposite (east) side of Granny White Pike about two miles distant." Soon they were in full retreat.

The Confederates fleeing southeast from Compton's Hill were trying to reach the escape route of the Franklin Pike two miles away. The closer route, Granny White Pike, was cut off by Federal cavalry, now intending to move northward up the pike. The trap was closing ever tighter and if not for the quick action of several elite units, Cheatham's Corps might have been bagged then and there.

The man of the hour was Brig. Gen. Daniel Harris Reynolds, leader of a brigade of Arkansans in Walthall's Division of Stewart's Corps. Two days previous, the Ohio native had celebrated his 32nd birthday. His father was from Virginia; his mother had ties to Maryland. He attended Ohio Wesleyan College and settled in Arkansas just before the war. He recruited a company and led the 1st Arkansas Cavalry (dismounted) at Richmond, Kentucky, so well in 1862 that his men were designated the first unit to march into Lexington.

Reynolds had been sent southward to assist Coleman about 3:30 pm. The men of Reynold's Brigade and Ector's Brigade under Coleman fought off the Federal cavalry and held Wilson's men west of the Granny White Pike. At 4:30 pm, Reynolds moved south to a gorge through the hills, about 500 yards east of Granny White Pike. He placed the 1st Arkansas at the north end and formed the rest of his brigade further south, and "threw up some rails for protection."

Ector's Brigade under Coleman was atop a high hill at the southwest side of the entrance to the gorge. The heavily wooded hills in this vicinity are very steep. Some Confederates managed to climb through the Overton Hills proper and make it to the Franklin Pike but most funneled through the gorge. Reynolds and Coleman maintained their positions for 30 minutes, allowing Cheatham's Corps to safely reach the Franklin Pike. Then they withdrew in order, fending off the pursuing cavalry in a night fight about 5:30 pm. Reynolds fell back and formed his brigade at the south end of the gorge. He then withdrew, covering the retreat as the rearguard. The country road through the gorge roughly follows present-day Overton Lea Road and Lakeview Drive. Federal division commander Cox later credited Reynolds and Coleman with saving a large part of Hood's army.

Robert H. Dacus of the 1st Arkansas Mounted Rifles, Reynolds' Brigade: "The Federals tried every means they could to force us out of this valley before dark. They had tried charging our front with infantry, and failing in this had tried flanking us with cavalry, then with infantry. Then they would try charging our front again."

"The Arkansas men fought a heroic series of rearguard actions on the Confederate left," said Gale, who found his commander, A.P. Stewart, along the Franklin Pike. The corps commander was surprised and glad to see him, having given up Gale for dead or captured. Stewart moved south to Brentwood, where he ordered engineer Wilbur Fisk Foster to lay out a line of defense.

## The Center and the Right Flank

The Confederates of Stewart's Corps east of Granny White Pike and west of Peach Orchard Hill hunkered down behind the Lea-family-farm stone wall beneath a storm of artillery fire from Federals on high ground directly in their front. The men of Major General William "Wing" Loring's division (the brigades of Featherston, Adams, and Scott) had caught hell the day before and had suffered significantly during the Battle of Franklin. In trenches farther east were the brigades of Deas and Manigault in Johnson's Division of Lee's Corps. The Confederates in the center were backed by the artillery batteries (west to east) of Bouanchaud (Point Coupee), Dent of Alabama, and Douglas of Texas. Despite their dire plight, the men of Stewart's Corps were surprised at noticing the avalanche of fleeing countrymen to their left coming from Compton's Hill. However, it took them little time to join in the hasty retreat as the Federal infantry of Wood's IV Corps

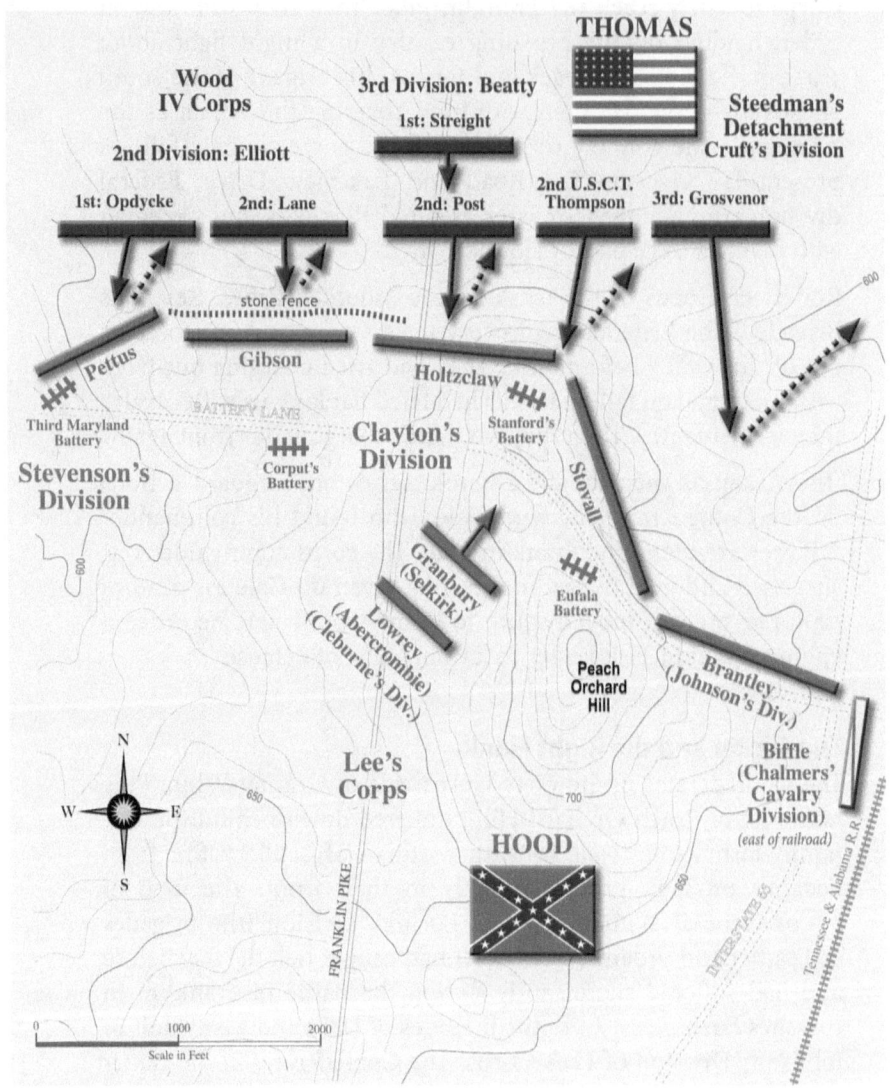

stormed their works. The collapse occurred so fast that most of the Confederate batteries lost guns. Horses had to be brought up and attached to the limbers, caissons, and artillery pieces in order to save them, and this procedure took time. Bouanchard lost all four of its artillery pieces. Dent's battery limbered up but got stuck in the mud and was captured before reaching the Franklin pike. Douglas' Texans lost one of their four guns.

The Confederates on the right flank at Peach Orchard Hill (a gentler slope than Compton's Hill) had enjoyed martial success most of Friday, December 16th, and therefore were caught off-guard by the sudden pandemonium on their left. The right-flank corps was commanded by Stephen D. Lee, the youngest lieutenant general in the Confederate army. Lee's Corps had arrived late at Franklin and avoided mass casualties, and the men had been largely unengaged the first day of battle at Nashville. Lee's Corps consisted of three divisions:

- Major General Carter Stevenson (brigades of Cummings and Pettus);
- Major General Henry Clayton (brigade commanders Brigadier Generals Marcellus Stovall, Randall Gibson, and James Holtzclaw); and
- Major General Edward Johnson (frankly, his four brigades did not fight well at Nashville).

Clayton, 37, of Georgia, a graduate of Emory and Henry College in Virginia, moved to Alabama and opened a law office. Two of his sons became U.S. Congressmen. Clayton served in the state house prior to the war. He recruited and organized a local militia and was elected as its captain. In August 1860, he was elected colonel of the 3rd Alabama, a statewide militia. Later, the governor authorized him to raise a new regiment for Confederate service, the 39th Alabama. Clayton was appointed its first colonel. Clayton's first significant campaign was the 1862 Kentucky campaign. He fought at Stones River, suffering a severe wound but recovered and was promoted to brigadier general in April 1863. He was assigned command of a brigade of Alabama regiments previously led by A.P. Stewart. His brigade played a prominent role in several fights during the 1864 Atlanta campaign. Following the Battle of New Hope Church, Clayton was promoted to major general and assigned command of

Stewart's old division in the Army of Tennessee.

Holtzclaw, 31, of Georgia, led Clayton's old brigade. He practiced law in Montgomery, Alabama, after declining an appointment to West Point. In early 1861, Holtzclaw served as a lieutenant in a local militia company, the Montgomery True Blues. He participated in the capture of the U.S. Navy Yard in Pensacola, Florida. He enlisted in the Confederate army as a lieutenant in the 18th Alabama Infantry. In August 1861, he was promoted to major and then in December to lieutenant colonel. He was severely wounded in the lung at Shiloh but made a remarkable recovery. He was promoted to colonel and served for a time in Montgomery. In 1863, he was again wounded, thrown from his horse at Chickamauga. In November 1863, he assumed command of a brigade in Stewart's Division and led it during the Chattanooga campaign. In July 1864, Holtzclaw was promoted to brigadier general.

S.D. Lee's men fought that day from fortified heights and repelled several vicious attacks by U.S. Colored Troops and others. Lee said his men "reserved their fire until (the enemy troops) were within easy range, and then delivered it with terrible effect." The 13th USCT took more casualties than any other regiment in the battle, losing nearly 40 percent of its strength. Six Federals were shot down trying to plant their battle flags atop Peach Orchard Hill.

Lee's Corps on the right flank was served by several artillery batteries: Stephens Light Artillery of Georgia (formerly the Third Maryland) sited 300 yards west of the hill; Van den Corput of Georgia with two guns; Stanford of Mississippi on the hill; and Eufala of Alabama on the east side of the hill.

The Confederates suffered mostly from Federal artillery. CSA 1st Lt. William L. Ritter of Stephens Light Artillery: "The horses could not effectually be sheltered from the enemy's battery on the right, and they were falling rapidly. I sent a man to Capt. Rowan to ask permission to move the horses out of the range of the enemy's shells, but he refused, saying he might need them at any moment. I again sent to him, telling him that if the horses were not removed in a short time, every one would be killed; he replied, 'Let the horses remain where they are and keep the gun teams full at the expense of the caisson teams.' About this time a shell passed through the head of one of the wheel horses, and then exploded, a piece of which cut my sabre in two and killed my saddle horse."

Ritter added: "At half past twelve, Captain Rowan was struck by a piece of shell. The piece of shell entered his body about four inches below the pit of his left shoulder, near the spine, ranging upward through the chest, killing him instantly."

Hours later, the left flank of the Confederate line broke and took the center with them. The right flank was in danger of being enveloped. Not only did S.D. Lee's men need to fall back quickly, they were needed to keep open the Franklin Pike escape route. Ritter's battery lost all four of their guns.

The Confederate division of Major General Edward Johnson (Deas and Manigault) broke and ran. Eccentric, but a man of great warmth, Johnson needed a walking stick due to a wound suffered in 1862. He was captured at Spotsylvania in May 1864 and had been released in October and transferred to the Western Theater. Now, Old Clubby formed his men into a hollow square to receive the Federal assault, but the formation quickly melted away. Weakened by his prior imprisonment and his leg wound, and being dismounted, Johnson was easily captured by the Federals.

The rapid withdrawal of A.P. Stewart's men took Lee's Corps by surprise. They did not have much time to act. Lee noted that the grayclad soldiers to the left "were flying to the rear in the wildest confusion, and the enemy following with enthusiastic cheers."

Thomas John Wood of the Federal IV Corps, which was stationed in the center of the Union line: "I at once ordered the whole corps to advance and assault the enemy's works, but the order was scarcely necessary. All had caught the inspiration, and officers of all grades and the men, each and every one, seemed to vie with each other in a generous rivalry and in the dash with which they assaulted the enemy's works." Wood's IV Corps claimed 14 pieces of artillery, 980 prisoners, two stands of colors, and thousands of small arms.

Pettus' Brigade of Alabama regiments was on the extreme left at Peach Orchard Hill, closest to the pandemonium. Capt. George E. Brewer, commander of the 46th Alabama, explained: "While dealing with those in front I heard near-by shots to my left and rear, and upon turning in that direction, to my amazement, I saw the enemy in possession of the battery to the left of the 30th Regiment on my immediate left, the enemy mingling in a hand-to-hand conflict with that regiment as they moved on down its flank and rear...I looked to the right and saw the 20th Alabama also getting out of their works. That left me no alternative but to order my regiment to about face and move out. Before all were out I saw General Pettus forcing the 20th back, so I countermanded my order. Before getting fairly settled down to work, the 20th left again for the rear. I repeated my command to about face and make to the rear. Part of both officers and men stayed, showing signs for surrender. I ran along the line begging them to follow me out, but most of those in the trenches stayed. The few who had been persuaded to leave joined me."

Stovall's Georgia brigade was on the extreme right of Peach Orchard Hill and, thus, the entire Confederate army. He was ordered to remain in position and await instructions. Not receiving orders and seeing the imminent danger of capture, he moved off on his own. "This was not done, however, I should perhaps state, until the whole army had given way and I was left alone and unsupported, with the entire force of the enemy closing in upon me...not a man of my brigade, so far as I could see, left the works without orders, and I was able to march them out in good order and save the battery which I supported."

As previously mentioned by Lt. Ritter, the artillery units atop Peach Orchard Hill experienced many troubles joining the retreat in good order, and 16 guns were lost to the enemy. Stanford's Mississippi battery was overrun. "Stanford's guns were run out of the works but in the mad rush the horses were stampeded and only enough were secured to carry off one caisson," stated Capt. Charles Fenner, commanding the battalion. Brigade commander Holtzclaw wrote that Stanford's battery was "so badly crippled as to be immovable, scarce a whole wheel remaining in its carriages, sustaining, without works, a fire from 18 of the enemy's guns for seven hours." Due in large part to the rapid loss of horses in the teams, the company lost all four guns and 12 men killed and wounded. Men weren't the only ones to panic. Although majestic, horses are not particularly intelligent, with brains the size of a walnut, and are prone to herd instinct and panic.

As the Confederate troops swarmed from Peach Orchard Hill with the Federals in hot pursuit, S.D. Lee rallied the men about a half mile south on the Franklin Pike near the Overton estate, using

Randall Gibson and his brigade of Louisianans as a rearguard, with Captain Douglas swinging a few pieces of field artillery hastily into position. Farther south on the pike, Clayton began reforming his division, now consisting of the brigades of Stovall and Holtzclaw.

S.D. Lee's aide, Louis F. Garrard, noted, "But for this action on the part of General Lee, the Federals, who were advancing on the left flank and rear of our army in a full run, would have been on all the troops of General Clayton's Division in the rear before they would have had knowledge of their approach or time to get out in any order…The rally enabled Clayton's Division to form a nucleus, and they, together with other Confederates, principally Lee's Corps, formed a line of battle. Right at the wheel of one piece of artillery I recollect a drummer stood, a mere boy, and beat a long roll in perfect time, without missing a note. The line of battle formed across the pike…was certainly a brilliant array of colors, and struck me as a rally of color bearers."

General Lee rode "right in the midst of the fugitives and seized a stand of colors from a color bearer and carried it on horseback, making himself a conspicuous object for the Federal infantry." Lee proclaimed: "Rally, men, for God's sake. This is the place for brave men to die." His example was inspiring. "He looked like a very god of war," recalled a member of his escort.

Lee asked Capt. Fenner if he had a serviceable battery. Fenner had just observed the Eufala Light Artillery (protected by the 39th Georgia) moving to the rear and pointed it out. "Bring that battery here at once," shouted Lee. Several guns were eventually assembled and, according to Lt. Colonel Llewellyn Hoxton, "were immediately placed in position and used with good effect to protecting the retreat of the army."

Historian Horn wrote: "Only Lee's prompt action in rushing to the rear of Stevenson's division and rallying the men served to create the impression of organized resistance and caused the on-rushing Federals to halt their advance, at least temporarily. This gave Clayton and some of Stevenson's division time to fall to the rear in good order and form a new line. The Federals were advancing on the run and would surely have cut Clayton off, but for the Lee-inspired rally east of the pike."

Participating in the pursuit south down the Franklin Pike were the U.S. Colored Troops who had assaulted Peach Orchard Hill. "The colored soldiers joined in the pursuit with as keen a zest as the most enthusiastic of the veterans," wrote L.G. Bennett and William Haigh, historians of the 36th Illinois. Private Benjamin T. Smith of the 51st Illinois stated, "The rebels who were captured by the colored troops gave up with a very bad grace, the officers especially, thinking it was a disgrace to be captured by niggers who perhaps were their former slaves."

Told by Hood that the Federals were nearing Brentwood, S.D. Lee abandoned his line across the Franklin Pike near the Overton house and hastened southward half a dozen miles, establishing the new rearguard line at Hollow Tree Gap, south of Brentwood and seven miles north of Franklin, at about 10:00 pm. Wood's IV Corps bivouacked for the night several miles short of the gap.

### The Battle at the Barricade

While Wilson's cavalry and the infantry of Schofield and A.J. Smith pursued the Confederates retreating from Compton's Hill southeastward toward the Franklin Pike, elements of blue and gray cavalry clashed on Granny White Pike about two-and-a-half miles south of the hill, and also later about three-quarters of a mile even farther south at what is now called the Battle at the Barricade. Although Federal cavalry had choked off Granny White Pike south of Compton's Hill and eliminated it as an escape route, Confederate forces farther south now attempted to prevent the Federals from using the pike to move even much farther south to its connection with Franklin Pike (near Hollow Tree Gap) in their attempt to block and then capture or destroy Hood's army.

Granny White Pike was named for a crusty old frontierswoman who had operated a popular tavern, the ruins of which still stood at the gap through the hills. Lucinda Wilson was a native of North Carolina who married teacher and militiaman Zachariah White, who was killed at the 1781 Battle of the Bluff against the Indians at Nashville. For 18 years, the widow lived in dire poverty in North Carolina caring for her two orphaned grandchildren. In 1800 at the age of 60, she loaded an oxen cart and headed for Nashville with the two children and an old slave named Uncle Zachary. It took Lucy three years to complete the arduous 800-mile mountainous

Stonework at entrance to Brentwood subdivision depicting the saber duel at the barricade.

journey, earning money by selling baked goods along the way. She bought 50 acres on the old Middle Franklin Turnpike for $300 and built a stagecoach inn that would become famous for its cooking, baked goods, fresh fruit, brandy, and applejack. Among the guests of Granny White were Sam Houston, Andrew Jackson, James K. Polk, and Thomas Hart Benton.

Now, at Nashville, it was time for the Confederate cavalry to make a stand on Granny White Pike. At that time, Major General Nathan Bedford Forrest was near Murfreesboro with two of his three divisions—Abe Buford's and Red Jackson's. His third division was stationed at Nashville, that of Brigadier General James R. Chalmers. Chalmers commanded two veteran officers, Colonel Edmund W. Rucker and Colonel David Campbell Kelley.

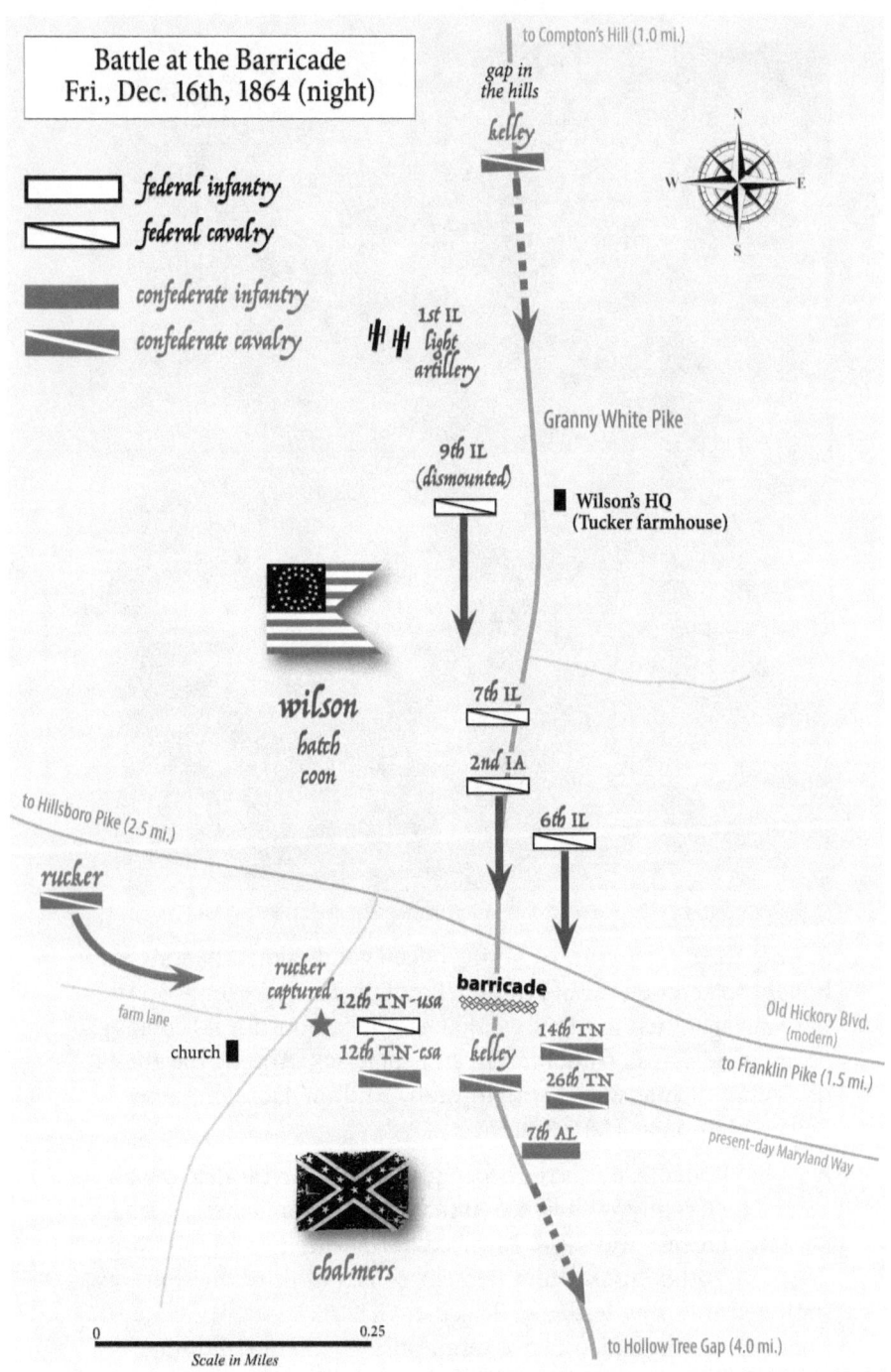

Mud, Blood & Cold Steel: The Retreat From Nashville-December 1864

Map drawn by Pvt. John Johnston of 14th Tennessee (CSA) during 1905 trip to battlefield.

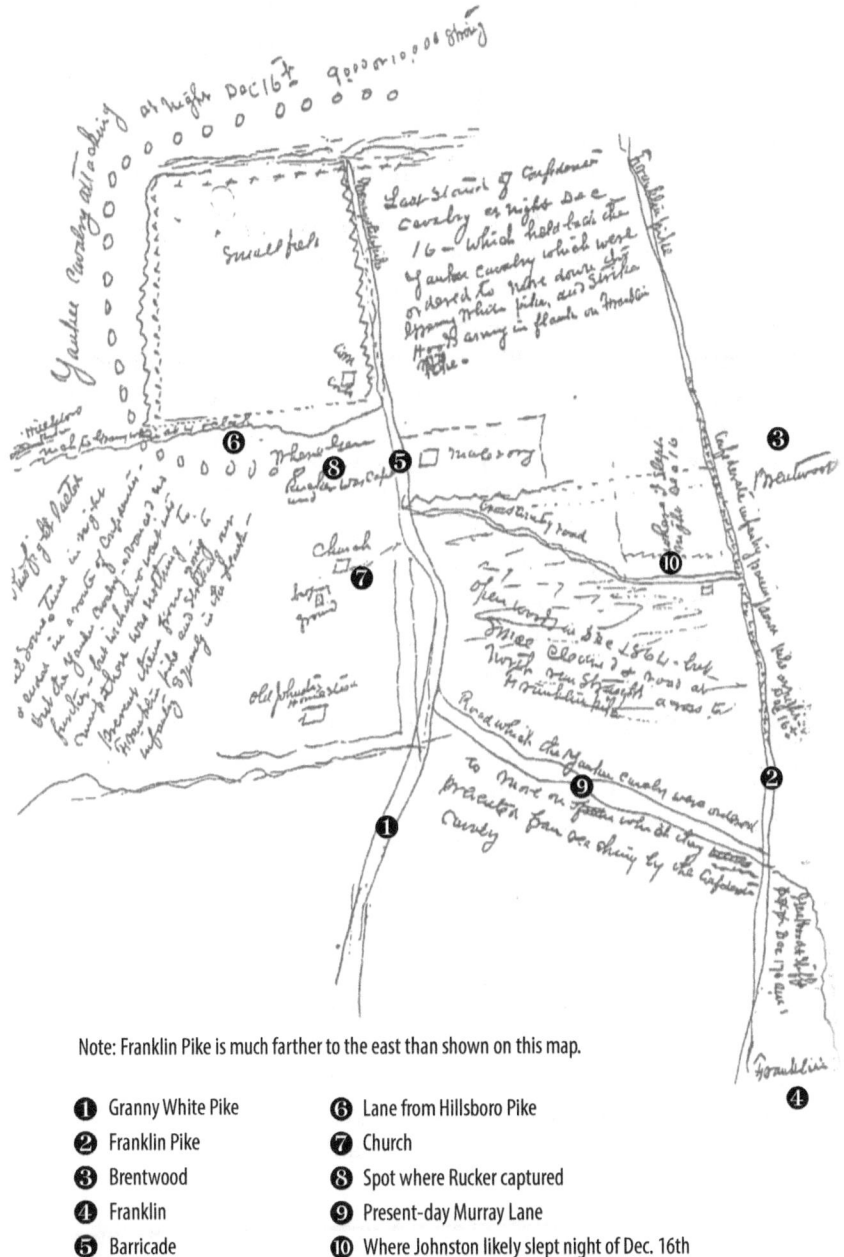

Note: Franklin Pike is much farther to the east than shown on this map.

① Granny White Pike
② Franklin Pike
③ Brentwood
④ Franklin
⑤ Barricade
⑥ Lane from Hillsboro Pike
⑦ Church
⑧ Spot where Rucker captured
⑨ Present-day Murray Lane
⑩ Where Johnston likely slept night of Dec. 16th

Chalmers, 33, was a native Virginian who spent most of his boyhood in Holly Springs, Mississippi, as the son of a U.S. Senator. By the time of the Civil War, he had practiced law, served as district attorney, and as a delegate to the convention which passed the ordinance of secession. Like his father, an ardent states' rights Democrat, Chalmers voted in favor of secession. One soldier described Chalmers as "a very genteel clever looking fellow."

Chalmers entered the Confederate States Army as a captain, and was soon promoted to colonel of the 9th Mississippi in 1861. In February 1862, he was promoted to brigadier general, and fought well at Shiloh, where he commanded Colonel N.B. Forrest. Chalmers tangled with Philip Sheridan as a cavalry commander at Rienzi, Miss. When Bragg advanced into Kentucky, Chalmers went along. His men fought well at Stones River. In April 1863, Chalmers was placed in command of the 5th Military District in the Department of Mississippi and East Louisiana. In 1864 he was assigned to command the 1st Division of Forrest's Cavalry Corps. This cavalry division was subsequently enlarged by the addition of Rucker's Brigade. General Chalmers played a conspicuous part in the Battle of Fort Pillow and in all the campaigns of Forrest in north Mississippi, West Tennessee, and Kentucky.

The relationship between Forrest and Chalmers was complicated. At the beginning of the war, Forrest fought under Chalmers; by the end of the war, Chalmers was subordinate to Forrest. Also, Chalmers was much more educated than Forrest, and occupied a higher rung in society. Friction occurred, but the two men managed to perform well in their duties.

During December 16th, Chalmers roamed to the south of Compton's Hill, attempting to fight off the advance of Wilson's troopers. He sent part of Rucker's brigade westward to the Hillsboro Pike. Then Chalmers took his escort to the county line at Brentwood, where the Confederate wagon train was parked, almost four miles south of the front line. As news of the debacle at Compton's Hill reached them, the teamsters began to drive their wagons rapidly southward on the pike.

At 4:30 pm, Chalmers received the urgent order from Hood to hold the Granny White Pike at all costs. Back on Hillsboro Pike, Rucker fought a brigade and artillery from Johnson's 6th Division for more than an hour. Outnumbered three-to-one, Rucker decided to join Kelley back on Granny White Pike, while the 7th Tennessee of his command moved down Hillsboro Pike, headed to Franklin to protect the supply wagons once they arrived there. D.C. Kelley's command tangled with Federal troopers on Granny White Pike about half-a-mile north of the county line and then fell back southward on the pike.

Reaching the Granny White Pike via a narrow farm lane, Rucker united with Kelley's men. Rucker's brigade, totaling 1,200 men, constructed a stout barricade of logs, brush, and fence rails across the roadway about 3.25 miles south of Compton's Hill, almost due west of the Brentwood rail station. Also manning the structure was the 7th Alabama, which included a company of cadets from the university at Tuscaloosa.

David Campbell Kelley possessed an interesting background. Known as the "Fighting Parson," he had served as a Methodist minister before the war. He was born in Leeville, Tennessee on Christmas Day 1833. An 1851 graduate of Cumberland University in Lebanon, Tenn., he became a medical doctor, graduating from the University of Nashville. That same year he traveled to China as a Methodist medical minister for two years. He began his Civil War service at Huntsville, Alabama, as captain of the Kelley Rangers, Co. F, Forrest's Battalion (3rd Tennessee Cavalry). He quickly became one of Forrest's intimates and most trusted associates. Kelley disdained the position of chaplain, according to biographer Michael R. Bradley, preferring to do his fighting as part of the "church militant."

Soon thousands of Federal cavalrymen under Hatch, Hammond, and Croxton arrived in the near darkness at the barricade. What resulted was the Battle at the Barricade, "one of the fiercest conflicts that ever took place in the Civil War," according to Wilson, who said his foe fought with "uncommon fierceness — a scene of pandemonium, in which flashing carbines, whistling bullets, bursting shells, and the imprecations of struggling men filled the air."

Above the din, while men fired their weapons at muzzle flashes, Kelley exhorted, "Pour it into them, boys, pour it into them."

During the melee, two prominent officers confronted each other on their mounts in a classic duel for the ages — Col. Edmund Rucker of the Confederate 12th Tennessee and Col. George Spalding of the Federal 12th Tennessee. Rucker was described by one Confederate soldier as "a man of great physical force and a fine horseman, he impressed men with his prowess in battle. Recklessly brave, he did not mind riding down an enemy, or engaging him in single combat. He helped to make the reputation of his old brigade as a body of fast and furious fighters."

Spalding, who served in Coon's Brigade of Hatch's Division, was a former school teacher from Michigan. The previous day, Spalding's men had captured Chalmers' headquarters wagons and prisoners at the Belle Meade plantation.

Reining in his white horse and clutching his saber, Rucker grabbed Spalding's bridle rein and declared, "You are my prisoner, for I am Colonel Ed Rucker of the Twelfth Tennessee Rebel Cavalry." Spalding replied, "Not by a damned sight!" and spurred his horse into action, breaking Rucker's hold. Captain Joseph C. Boyer of the 12th Tennessee (U.S.) wrenched Rucker's saber from his hand, and Rucker in turn grabbed Boyer's saber. They dueled and wrestled until a shot from a Federal trooper broke Rucker's sword arm. Rucker surrendered.

The grievously wounded Rucker was now Hatch's prisoner. His arm was mangled so badly that eventually it had to be amputated by a Federal surgeon. Rucker and Hatch ended up spending the night at the Tucker farmhouse near the pike. During their conversations, Rucker convinced Hatch (and subsequently Wilson) that Forrest was in the vicinity with the remainder of the Confederate cavalry and "will give you hell tonight." The rumor, although false, was enough for Wilson to call off any pursuit for the rest of the night. Besides, the Federal troopers had been maneuvering and fighting for 18 hours (plus the previous day) and needed a rest.

Of that night, Rucker later wrote: "General Hatch laid down on the floor by my side, and (God bless him) got up frequently during the night, and gave me water, and the next morning, when (the guards and I) left for Nashville, he provided me with a small flask of good whiskey."

Pvt. John Johnston of the 14th Tennessee vividly described the Battle at the Barricade: "We could see nothing; the mist and darkness had covered all in front, and we shot blindly out into the darkened woods. Our whole line from right to left was one continuous blaze and roar of musketry. How long this continued, I don't know. I thought about 30 minutes, but the historians and military men, who have written of this affair, say — some, about two hours, until near midnight; but I know that all came to an end very suddenly and unexpectedly."

Johnston added that, in his opinion, although Rucker was in command and entitled to some glory, to the men of the 14th Tennessee the gallant D.C. Kelley was the real hero. The private claimed that the men believed Rucker to be drunk and Kelley the de facto leader at the barricade.

Wilson later wrote that Chalmers and his cavalry had been "overborne and driven back, it is true, but the delay which he forced upon the Federal cavalry by the stand he had made was sufficient to enable the fleeing Confederate infantry to sweep by the danger-point that night, to improvise a rearguard, and to make good their retreat the next day."

Following the skirmish, the 14th Tennessee was ordered to form a picket line while Rucker's Brigade moved over to the Franklin Pike. The rearguard of the Army of Tennessee was reformed at Hollow Tree Gap (now called Holly Tree Gap) on the Franklin Pike where the Granny White Pike (meandering in a southeasterly direction) finally intersected, a little more than four miles north of Franklin. This congregation consisted of S.D. Lee, along with Stevenson's Division and part of Clayton's Division.

Beatty's Infantry Division of Wood's Federal IV Corps advanced down the Franklin Pike to within two miles of Brentwood. By this time it was dark and further pursuit was abandoned. A weary Federal soldier of the 64th Ohio noted: "As night was coming on, the 64th was ordered on picket. Our position fell in a corn-field. As is usually the case after a great battle, we had a pouring rain. Our condition was anything but agreeable in this mud and soaking rain. Tired and hungry we had to grin and bear it, very thankful that we were yet alive. Company D had 47 men when we met the enemy at Spring Hill, a little over two weeks ago. Now there were eleven of us to stack arms."

Federal cavalry commander Wilson spent much of the night of December 16th in the Tucker farmhouse, which doubled as headquarters and hospital. Thomas had caught up to Wilson on horseback. "Dang it to hell, Wilson, didn't I tell you we could lick 'em, didn't I tell you we could lick 'em?" Thomas wired Halleck, his chief in Washington, the next day and recommended Wilson for the full rank of major general of volunteers "for the excellent management of his corps during the present campaign."

Grant telegraphed Thomas at 11:30 pm to congratulate him "on your splendid success of today." Advocating total victory, Grant encouraged Thomas not to be content with merely stopping Hood's advance. "Push the enemy now and give him no rest until he is entirely destroyed," Grant ordered. "Much is now expected."

The pursuit would continue the next day, but the two-day Battle of Nashville was over. Federal casualties totaled 3,057, with less than 400 killed. On the second day of battle, the Federals captured 3,300 prisoners for a two-day total of 4,462. One-third walked without shoes. A Federal officer noted: "They were all ragged and dirty, and so filthy that we could smell them." Thirty-seven pieces of Confederate artillery were claimed, for a total of 53, along with 3,034 small arms.

Speculation has been waged by Civil War scholars about whether Wilson's troopers could have bagged Hood's army the night of December 16th. Historian Hay stated that Wilson, who commanded four divisions, could have succeeded. "…it is reasonable to contend that sufficient comparatively fresh detachments from these (four) divisions could have been quickly made up to have continued the pursuit. Only parts of Hatch's and Knipe's divisions were seriously engaged with Chalmers."

Hay added, however, "Though it fought dismounted, the Federal cavalry did more actual fighting in the battle of Nashville than on any other battlefield of the war. Instead of dispersing it on both flanks of the army the cavalry was concentrated on one flank where it could be of most decisive use. Under the conditions as they finally came to pass, a withdrawal under fire was hazardous and well-nigh impossible of successful execution, and when it was finally decided on, it was too late."

In 1896, General Henry V. Boynton, a Medal of Honor recipient for Civil War service, contended: "In no other battle of the war did cavalry play such a prominent part as in that at Nashville.

Unquestionably, Wilson's cavalry was the dominating and controlling element of the battle."

Confederate Private Johnston begged to disagree: "The truth is that the Federal armies were handled with a great deal of timidity, and this fine writing about the greatness of their achievements is all bosh."

Hood did not file an official report on the Battle of Nashville until two months later. Regarding the action at Compton's Hill: "Up to this time no battle ever progressed more favorably; the troops in excellent spirits, waving their colors and bidding defiance to the enemy. The position gained by the enemy being such as to enfilade our line caused in a few moments our entire line to give way, and our troops to retreat rapidly down the pike in direction of Franklin, most of them, I regret to say, in great confusion, all efforts to re-form them being fruitless. Our loss in artillery was great—fifty-four guns. Thinking it impossible for the enemy to break our line, the horses were sent to the rear for safety, and the giving way of the line so sudden that it was not possible to bring forward the horses to move the guns which had been placed in position. Our loss in killed and wounded was small."

### Forrest at Murfreesboro

Late on Thursday, December 15th, two dozen miles to the southeast, news of the opening of battle at Nashville reached Confederate cavalry commander Nathan Bedford Forrest at Murfreesboro and forced him to cancel plans to strike the Federal forage train east of Fortress Rosecrans. Instead, he moved his forces to Wilkinson's Cross Roads, six miles west of Murfreesboro. Late the next day, December 16th, he received word of the defeat at Nashville and was ordered by Hood to fall back via Shelbyville and Pulaski. Because he was already at Triune with his 400 prisoners and several hundred hogs and cattle, he directed his wagons more sharply to the west, through the rain and sleet, toward Columbia on the Duck River.

Forrest ordered his trusted division commander, Abe Buford, to fall back from the Cumberland River via La Vergne to the Nashville Pike, then shifted him to the Franklin Pike to serve in the rearguard of Hood's army. Later, he sent Brig. Gen. Frank

Armstrong there for the same purpose.

Buford's Division frequently functioned as two distinct cavalry units, Buford commanding the five regiments of Kentuckians in person, and Colonel Tyree Harris Bell commanding the five regiments of Tennesseans.

Abe Buford, 44, was born in Kentucky and named for his great-uncle Abraham, who was a Continental Army officer during the Revolutionary War. He was descended from a Huguenot family named Beaufort, which fled persecution in France and settled in England before emigrating to America in 1635. His cousins, John and Napoleon Bonaparte Buford, were Federal generals.

Mercer Otey, Forrest's adjutant, described Buford: "He weighed something over three hundred pounds, of powerful frame, a round ruddy face covered with a short, stubby red beard, dressed in brown butternut Kentucky jeans, his pants invariable stuck in his boots…With all of his weight, he was the most graceful dancer I ever saw swing a lady…"

Buford also had an eye for horseflesh. An 1841 graduate of West Point, Buford fought on the frontier and in Mexico, where he was cited for bravery. Following the outbreak of the Civil War, like his native state, Buford tried to stay out of the fighting and succeeded in doing so for well over a year. In September 1862, during Bragg's invasion of Kentucky, Buford joined the Confederate army. He helped raise and took command of a Kentucky brigade and in September 1862 was commissioned brigadier general. Buford covered Bragg's retreat from Kentucky, served in the Vicksburg campaign, and fought at Champion Hill. He raided Paducah, Ky., in March 1864 under Forrest, and fought at the Battle of Brice's Crossroads.

Tyree Bell was older at 49, had led infantry at Shiloh, and was severely wounded during that battle. Promoted to colonel, he led his regiment during the invasion of Kentucky and the battle at Richmond. Joining Forrest's command, Bell became one of his most trusted subordinates. In 1864, Bell and his brigade fought at Fort Pillow, Brice's Crossroads, and against A.J. Smith's force in Mississippi in August 1864 after the Battle of Tupelo. Bell was wounded in the chest, back, and face at Pulaski, Tennessee, on Sept. 27, 1864. He continued to serve under Forrest and led his brigade at Johnsonville in November 1864.

Bell made his way on December 16th from Murfreesboro up the pike to Nashville and then crossed Nolensville Pike to the Franklin Pike, with Major Tom Allison guiding. They came to the Franklin Pike north of Hollow Tree Gap at 2:00 am the morning of December 17th. Bell sent Colonel John F. Newsom of the 19th Tennessee about a mile north up the pike to the Isola Bella plantation.

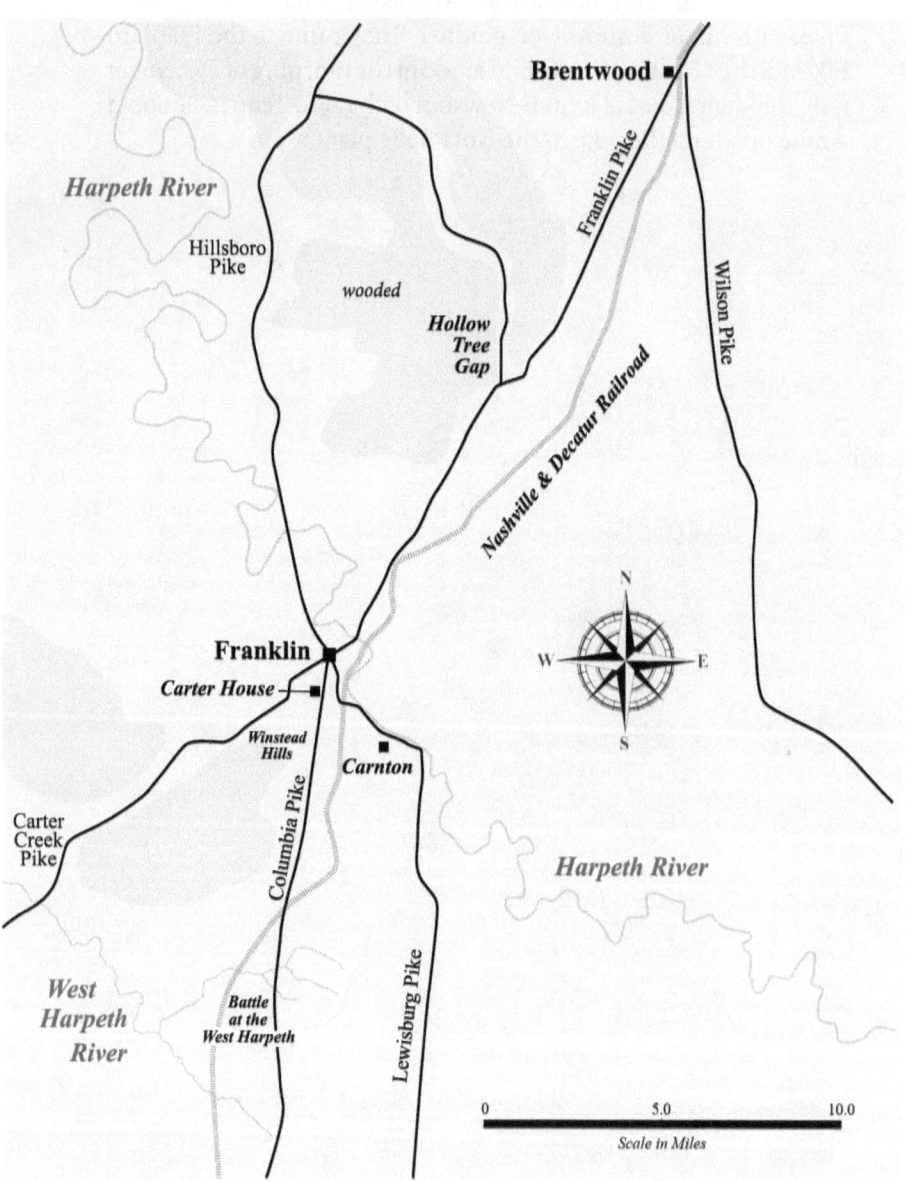

## Hell-Bent For Leather
*Saturday, December 17th, 1864*

There wasn't much rest for the weary that night, as the shattered Army of Tennessee made its way past Brentwood and southward to Franklin, remnants of the same army that had marched boldly northward into Tennessee just three short weeks before. The disheveled soldiers marched 12 miles on the pike that night, south to the Harpeth River and Franklin, most of them reaching the river by 3:00 am, with the rearguard stationed four miles back at Hollow Tree Gap.

What the men found at Franklin was unnerving and disheartening. The town, site of fierce fighting two weeks earlier, sheltered more than 2,000 wounded, including 250 Federals, in 45 makeshift hospitals converted from commercial buildings, churches, and even houses.

Pvt. Stephenson said, "Many of the wounded of Franklin had joined us as we fell back through the town, refusing to be captured and now limped or struggled along as best they could by our sides, still dauntless and defiant." As the members of Co. H, 9th Tennessee, slogged through the small town, a private on cook detail handed each one a piece of half-baked bread, soft from the rain. Reportedly, it was their first food in 24 hours and it would have to satisfy them for the next 36 hours.

Lieutenant R.M. Collins of the 15th Texas: "I kept no company that night except that of my (confiscated) mule. On we went, but could find no good place to sleep. After a while we came to Franklin. We crossed the creek, passed on through the city, by the battle ground, and a way out south of the city found a battery camped…I slid down off my mule, tied him, spread my blankets, crawled under a caisson and fell asleep."

Back at Nashville, the ghastly but necessary job of clearing the battlefield, tending to the wounded, and burying the dead — man and beast — began for the victorious Federals. At the Compton farmhouse, approximately 150 dead and wounded Southern soldiers were laid out, the wounded tended to as much as possible. Virginia Cliffe, the wife of regimental surgeon Dr. Daniel Cliffe, recovered the body of Colonel William Shy atop Compton's Hill. Shy was laid out on the front gallery of the farmhouse with "a cruel hole in his brow," according to Felix Compton's daughter. Later, Shy would be buried in the family graveyard at Two Rivers on Del Rio Pike in Franklin.

At his temporary headquarters, Federal cavalry commander Harry Wilson worried about the orders he received from Thomas at 3:00 am. The chief wanted Johnson's cavalry division to remain on the Hillsboro Pike to the west while the remainder of the Federal cavalry would move east to the Franklin Pike. Wilson thought it better to attack down the Granny White Pike since it eventually terminated at the Franklin Pike at Hollow Tree Gap. But orders were orders. He sent Knipe and Croxton to the Franklin Pike (Knipe was responsible for the pike and those roads west of it; Croxton the roads east of it, eventually moving south down the Wilson Pike) and then he and Hatch followed suit. The troopers had one purpose in mind — cut off Hood before he reached the Harpeth River, immediately north of Franklin. They could not wait for the infantry, which had to use the macadamized pike. Delayed by rain and darkness, the Federal IV Corps infantry under Thomas John Wood did not begin movement down Franklin Pike until 8:00 am. Of course, Wilson did not know that the main body of the Army of Tennessee had already crossed the Harpeth River at Franklin.

Wilson wanted Johnson to move down the Hillsboro Pike to Franklin and hit Chalmers, who was protecting Hood's wagon train, in the flank. "Give him no peace," Wilson told Johnson. "Time is all he wants. Don't give him any."

Wilson was chomping at the bit to get started in his quest to bag Hood's army. This late in the year, the sun rose at 7:00 am and provided only about nine hours of daylight. The steady rainfall was turning all the terrain off the turnpike into a muddy quagmire and raising the level of the waterways so fast that few places were still fordable. The Confederates, of course, were destroying the bridges after crossing the rivers.

At this point, the Federal cavalry corps consisted of 13,910 men, plus 338 in the headquarters unit. They were organized as follows (each division included a battery of light artillery):

- Croxton's brigade of 1,388 troopers;
- Hatch's 5th Division of Stewart's 1st Brigade and Coon's 2nd Brigade, totaling 4,621 men;
- Richard Johnson's 6th Division of Harrison's 1st Brigade and Biddle's 2nd Brigade, totaling 4,034 men; and
- Knipe's 7th Division of Hammond's 1st Brigade and Gilbert Johnson's 2nd Brigade, totaling 3,867.

A short ways out the Franklin Pike, Hammond's troopers came upon the Confederate rearguard of Col. John Newsom's 19th Tennessee below Brentwood and drove them back to Hollow Tree Gap. The Federals found Pettus' brigade stationed southwest of the gap with Stovall's brigade across the pike. Bledsoe's Missouri artillery battery was deployed just south of the gap, with Holtzclaw's and Gibson's brigades in reserve. Among the units in reserve on the right flank of the pike were the 4th and 30th Louisiana.

At 9:00 am, at Hollow Tree Gap, two mounted Federal regiments attempted to overpower the rearguard. While the 19th Pennsylvania drove down the pike, the 10th Indiana circled around the Confederate right flank and threatened to engulf it. The Federals were initially repulsed with the loss of 22 killed and wounded and 63 captured. But then the rearguard, worried that the Hoosiers might outflank them, fell back toward Franklin. The 10th Indiana Cavalry, under Lt. Colonel Gresham, captured two flags, two colonels, two lieutenant-colonels, and 110 enlisted men, mostly from the 4th Louisiana and 30th Louisiana under the command of Colonel Samuel E. Hunter.

General S.D. Lee reported: "Early on the morning of the 17th our cavalry was driven in confusion by the enemy, who at once commenced a most vigorous pursuit, his cavalry charging at every opportunity and in the most daring manner." Clayton said that Chalmers' cavalry was driven back in a "most shameful manner" as the Federals cut with their sabers, right and left.

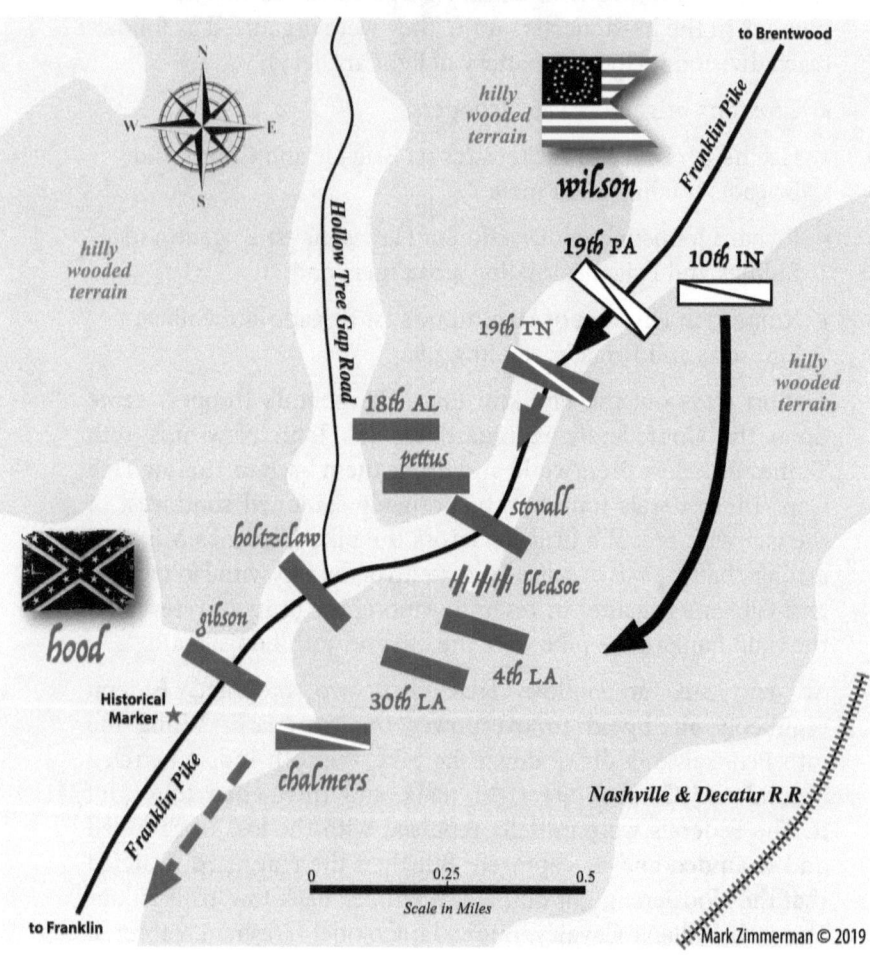

Lee's rearguard had been bested at Hollow Tree Gap, but his stand bought Hood and much of the infantry enough time to cross the Harpeth River at Franklin. Nathaniel Cheairs Hughes Jr., biographer of Confederate Colonel Tyree Bell, wrote of Hollow Tree Gap: "This small cavalry rear-guard action marked the beginning of the systematic, effective deployment of cavalry as a defensive screen for Hood's retreating army. The Federal cavalry from this day forward would exercise far more caution in their pursuit, worrying about rushing into an ambush as they attempted to bring Hood's army to bay."

The 9th Indiana cavalry was bringing up the rear, observing all the debris and accoutrements abandoned by the fleeing Confederates. The Federals concluded from such evidence that their pursuit would be "a walk-over." Capt. Obediah Hayden of the 9th Indiana ruefully noted later that "Hollow Tree Gap undeceived us."

Capt. Hayden chronicled one of the many horrors of war as his unit moved down the pike: "A trooper lay beside the road gasping his life away, and near him with a ghastly wound in his breast, lay dead the little curly-headed, blue-eyed boy, Duane A. Lewis, Co. B., sixteen years old, the General's orderly, whose bright and joyous face and fearless innocence had endeared him to the heart of every soldier in the brigade. The pitiless rain fell upon his upturned childish face; his eyes were open, but their light had gone out forever."

It should be noted that during much of the retreat that day the fighting was nearly continuous, with small skirmishes breaking out on a routine basis. Nevertheless, the next significant action took place just north of Franklin at the Harpeth River. In the words of the historical marker now there, "perhaps the largest cavalry engagement on American soil took place along Franklin Pike and the Nashville & Decatur RR sweeping across what is now Harlinsdale Farm." (The Battle of Brandy Station in Virginia, fought on June 9, 1863, with 18,456 horsemen involved, is generally acknowledged as the largest cavalry battle of the war.) Wilson's cavalry pursued with the thundering of thousands of hooves, the main column advancing down the Franklin Pike, with Hammond's troopers closing in from the east on Liberty Pike. Waiting there before the river was Buford's cavalry division, Pettus' brigade, the 13th, 16th, and 19th Louisiana regiments commanded by Gibson, and Bledsoe's artillery battery, which was arrayed along Front Street across the river.

Randall Gibson, 32, was a former member of the U.S. House of Representatives and U.S. Senator from Louisiana. His great-great-grandfather was a free man of color who was married to a white woman, and had owned land and a few slaves in Virginia, before migrating with other settlers to South Carolina in the 1730s. Gibson's father moved his family to Louisiana when Randall was a child. He graduated from Yale University in 1853, and returned home to study for his bachelor of laws from the University of Louisiana Law School, later Tulane University. Soon after Louisiana's secession, Gibson became an aide to the governor. In May 1861, he left the capital to join the 1st Louisiana Artillery as a captain. In August 1861, he was commissioned as colonel of the 13th Louisiana. Gibson fought at Shiloh and subsequent actions. With the Army of the Mississippi, he took part in the 1862 Kentucky campaign and the Battle of Chickamauga. After being promoted to brigadier general in January 1864, he fought in the Atlanta campaign. Three weeks previous, at the beginning of Hood's Tennessee campaign, Gibson's men had been the first to enter Florence, Alabama, having paddled across the Tennessee River in pontoon boats under heavy fire.

The 250 men of Gibson's brigade manned a small redoubt 1,000 yards in front of the Harpeth River and the pontoon bridge across it. His unit now comprised less than one-tenth their number at the beginning of the war. This despite the fact that they missed the battle at Franklin and were lightly engaged on the first day at Nashville. Diversity was their lot. The 20th Regiment was composed of German and Irish immigrants. The 13th Regiment from New Orleans consisted of "Frenchmen, Spaniards, Mexicans, Dagoes, Germans, Chinese, Irishmen and, in fact, persons of every clime known to geographers or travelers of that day." Other regiments consisted of wealthy gentlemen, young sugar planters, and slaveowners. The brigade also included Austin's battalion of sharpshooters. They all had served well as part of the rearguard on December 16th.

At 10:30 am, Wilson's cavalry of 3,000 men crashed into the Confederate earthworks with a cacophony of horses whining and stomping, yells and screams, and concussive explosions. Vastly outnumbered, Gibson's men and the supporting cavalry fought as long as they could before being nearly surrounded. They then fled over the pontoon bridge, having lost 13 killed, 25 wounded, and 364 captured.

Wilson later recalled: "It was killing work for both sides. The rain was still pouring and the fields on both sides of the road were soaking wet." Among those lost, according to Hammond, was Captain Hobson of the 9th Indiana Cavalry, "a man remarkable for the prompt discharge of his duties and his bravery. He is a great loss to the service."

Capt. Hayden of the 9th Indiana cavalry described the ill-advised Federal charge upon the earthworks. The Confederates had torn down telegraph wire and, driving posts at intervals, had encircled the fort with it. This was not noticed by the Federals until their horses tumbled over it. The lieutenant of Company C had part of his skull torn away by a fragment from a bursting shell. Another shell passed through two horses, taking off the leg of one of the riders. Another horse had his head taken off "as with a broad-axe." During the charge, a horse was struck full in the chest with a cannon-ball, passing through and disemboweling him. The rider went headlong in the mud, where he lay stunned until the fight was over. The 9th Indiana fell back in disorder but not before capturing two stands of colors and 200 prisoners. The prisoners were taken by individual prowess and were not the result of concerted action. The regiment fell back 200 yards and reformed.

While all this bedlam transpired, S.D. Lee's engineer, Capt. Coleman, and his pioneers destroyed the trestle railroad bridge over the Harpeth River under fire of Federal sharpshooters. Nearby, Confederate pioneers in Lee's Corps worked quickly to destroy the pontoon bridge. Aiding the Confederates in their escape was the section of Bledsoe's Battery posted along Front Street in Franklin. Their shells exploded above the heads of Wilson's cavalrymen, causing them to hesitate in their onslaught. During this time or perhaps earlier, the Confederates, in their desperation to lighten their load and flee, dumped thousands of lead bullets into a depression and covered them with a large cast-iron camp kettle. The cache, weighing a total of 1,250 pounds, was discovered 150 years later by relic hunter Williamson Henry and can be viewed today at the Lotz House Museum on Columbia Pike.

Galloping down the Hillsboro Pike, Richard Johnson's 6th Division of U.S. cavalry forded the Harpeth River west of Franklin and reached the village about the same time as the other Federal troopers, having taken more than six hours to advance nine miles down the pike. Assigned to cut off Hood's fleeing army, the troopers instead watched as the Confederates flooded the streets of Franklin. Harrison's troopers met a detachment of Bell's cavalry and "we go for him at once and soon move him out," according to a trooper of the 7th Ohio. Forty prisoners were taken in the process. Knipe's men forded the river near the demolished railroad bridge, taking prisoner some 75 men from Holtzclaw's brigade who had been left behind.

Major Eugene F. Falconnet, commanding the 7th Alabama cavalry, was about to cross the Harpeth River when he noticed the enemy charging upon Gibson's brigade. He gathered about 100 followers, drew his revolver, and charged the enemy, more than 20 times his number, apparently gaining Gibson some more time to retreat.

The Harpeth River at Franklin was a formidable obstacle for the Federals in their pursuit, only the first of many rivers and overflowing streams they were to encounter. Wood's IV Corps of infantry, moving down the main pike, was delayed for 18 hours at the Harpeth. Wood recalled: "Colonel (Isaac) Suman, Ninth Indiana, nobly volunteered to build the bridge, and thanks to his energy and ingenuity and the industry of his gallant regiment, it was ready—though he had few conveniences in the way of tools, the scantiest materials, and the stream was rising rapidly—for the corps at daylight the morning of the 18th."

It was becoming apparent to the Federal infantry that the best of their pontoniers were away with Sherman in Georgia. The Federal pontoon train was desperately needed at the Harpeth River, but where was it?

Following victory at Nashville, during the early morning hours of December 17th, General Thomas awoke from a deep sleep and gave the orders to move out the pontoon train as soon as possible. Major James B. Willett, in charge of the pontoons, was told to move the pontoon train as early as possible, down the Murfreesboro Pike. Thomas' adjutant, Capt. Robert H. Ramsey, actually wrote the order. Willett marched the train 15 miles down the Murfreesboro Pike before the orders were changed to the Franklin Pike. Willett backtracked and then took an impassable side road to make up time, but his 500 horses and mules got mired down in the mud. He was forced to backtrack all the way to Nashville and move out on Monday, December 19th. Due to this critical mistake, the pontoons failed to arrive at the front of the army until December 21st, five days late.

Apparently, Thomas also had spoken to Capt. William LeBaron Jenny, an engineer, who tried to explain the debacle. "After the battle (at Nashville), General Thomas sent for me, and told me to take my train out on the Murfreesboro pike. I said, 'General, do you mean the Murfreesboro Pike?' because I knew that was away from the enemy. He said 'Yes, on the Murfreesboro Pike.' I went away, but I was uneasy in my mind, for I knew that a bridge train could not be wanted where there were no rivers. I turned and went back, and again I asked him, 'General, do you mean the Murfreesboro Pike?' He seemed heavy, but aroused himself, and half-angrily said, 'Yes, the Murfreesboro Pike, go and execute your orders.' I went and led out my pontoon, as directed."

Moving all the components of a pontoon bridge was a major undertaking. A typical pontoon-boat wagon train (Federal) consisted of 34 pontoon wagons, 21 chess (bridge floor) wagons, 4 tool wagons, and 2 forge wagons, according to Rice's logistics reference. The train employed a minimum of 372 horses or mules, not accounting for replacements. For a 10-day routine march, these animals would consume 96,720 pounds of forage, which requires a supply train of at least another 33 wagons and the 198 horses to pull them.

As the Confederate troops moved through Franklin and then onto the former battlefield of two weeks prior, they witnessed much suffering in the 40 or so makeshift hospitals. They grimly examined the detritus of the recent battleground. A corporal in the 18th Alabama stated: "We marched along over these (burial) ditches, our feet sinking down in the soft, wet dirt. I saw more than one hand or foot exposed and I distinctly felt my foot rest on a dead body as it sunk into the mud. It was a gruesome sight. And then the stench arising from these dead bodies was something fearful."

Hard luck haunted one Confederate general officer languishing in Franklin. Brig. Gen. William A. Quarles, 39, a Tennessee lawyer and politician before the war, led a brigade in Walthall's Division of Stewart's Corps at Franklin on November 30th. He already had been captured in 1862 at Fort Donelson and served time in a POW camp before being exchanged. Then he was severely wounded at Pickett's Mill in Georgia. Then at Franklin a shell shattered his left arm. Now hospitalized in Franklin and too ill to move, Quarles was again taken prisoner as the Federals moved through and continued their pursuit. In the lead were Col. Israel Garrard's 7th Ohio of Johnson's cavalry and Hammond's 4th Tennessee commanded by Lt. Col. Jacob M. Thornburgh. Soon, Franklin was once again in Federal hands.

Determined not to be captured, another patient in Franklin, E.H. Wingate of the Washington Artillery, who had been shot through the chest at Overall Creek, was able to leave his bed and rejoin the Army of Tennessee as it passed through.

Union loyalist Fannie Courtney, 19, whose home was two blocks south of the Franklin square, wrote: "What a happy set of men the Federal wounded were when they heard the glad news! It was on Saturday, December 17th, when the advance cavalry of our troops entered the town. I was at the hospital. What shouts were given by those who were able to creep to the door!"

Private W.A. Keesy of the 64th Ohio noted: "On the 17th we came to Franklin. I was anxious to find some of our missing boys. I stole away from the ranks and made for the hospital. The store buildings were turned into hospitals and were filled with both Federal and Confederate wounded. On coming to the first hospital I was confronted on the porch with men terribly wounded… One man, shot through the jaw, his tongue protruding out of his mouth, rested his head on his hands. He could not speak. Another, who was shot in the thigh but able to be laid out on the porch, was badly doubled up."

Stepping onto the battleground from two weeks prior, Pvt. Keesy was haunted by recent memories: "As far as my eyes could see there were rows of graves, side by side, where mostly men who had died of wounds since the battle were buried. The partially filled ditches told where the dead of the battle were laid. As I stood there and thought of the awful suffering and slaughter of the battle, and how nearly I had come to being one of the number to inhabit those ditches, I trembled; and from my heart I thanked God, and fled from the spot to join my comrades."

In the general scheme of events, however, there was very little time for reflection. It was rough work, retreating through the town square, then due south on the pike. Colonel Bell of the Tennessee cavalry stated, "We fought nearly constantly the whole day and until some time in the dark, alternately fighting and falling back. The enemy pressed us hard…late in the afternoon they run in on us and got Buford cut off. The general (Buford) being on a better saddle horse than most of us, and being of such a powerful build, he knocked the federal down that attacked him with the butt of his pistol and leaped a ditch with his horse."

South of Franklin, the pike continued through rolling countryside to the sizeable town of Columbia in Maury County on the Duck River, about 25 miles away. To get there, the Confederate soldiers, pursued by the Federal cavalry and infantry, would need to cross through the Winstead Hills and over the West Harpeth River, pass through Thompson's Station and Spring Hill, and then cross over Rutherford Creek. The last obstacle to reaching Columbia would be the Duck River itself.

At 1:30 pm, just south of Franklin, Federal cavalry commander Wilson reorganized his troopers. Richard Johnson's division would continue to move along the Federal right flank, down Carter's Creek Pike to the west. Hatch and Knipe would pursue directly down what was now the Columbia Pike in two parallel columns. Croxton's brigade would advance eastward along the Lewisburg Pike, cross the Harpeth at McGavock's Ford, and eventually halt for the night of December 17th at Douglas Church. Along the way, Croxton would capture 130 more prisoners. The Federal cavalry would advance with no infantry support that day — Wood's IV Corps was stalled at the Harpeth River in Franklin.

Brigadier General John T. Croxton, 28, was a Kentucky native and Yale graduate. He practiced law in the Bluegrass State and became lieutenant colonel of the 4th Kentucky Mounted Infantry. He was severely wounded at Chickamauga, and elevated to brigadier general in July 1864. He commanded a cavalry brigade in the Atlanta campaign. He was tall, dark-eyed, and handsome, an ardent abolitionist, and an enthusiastic soldier.

About this time, Confederate Major General S.D. Lee personally

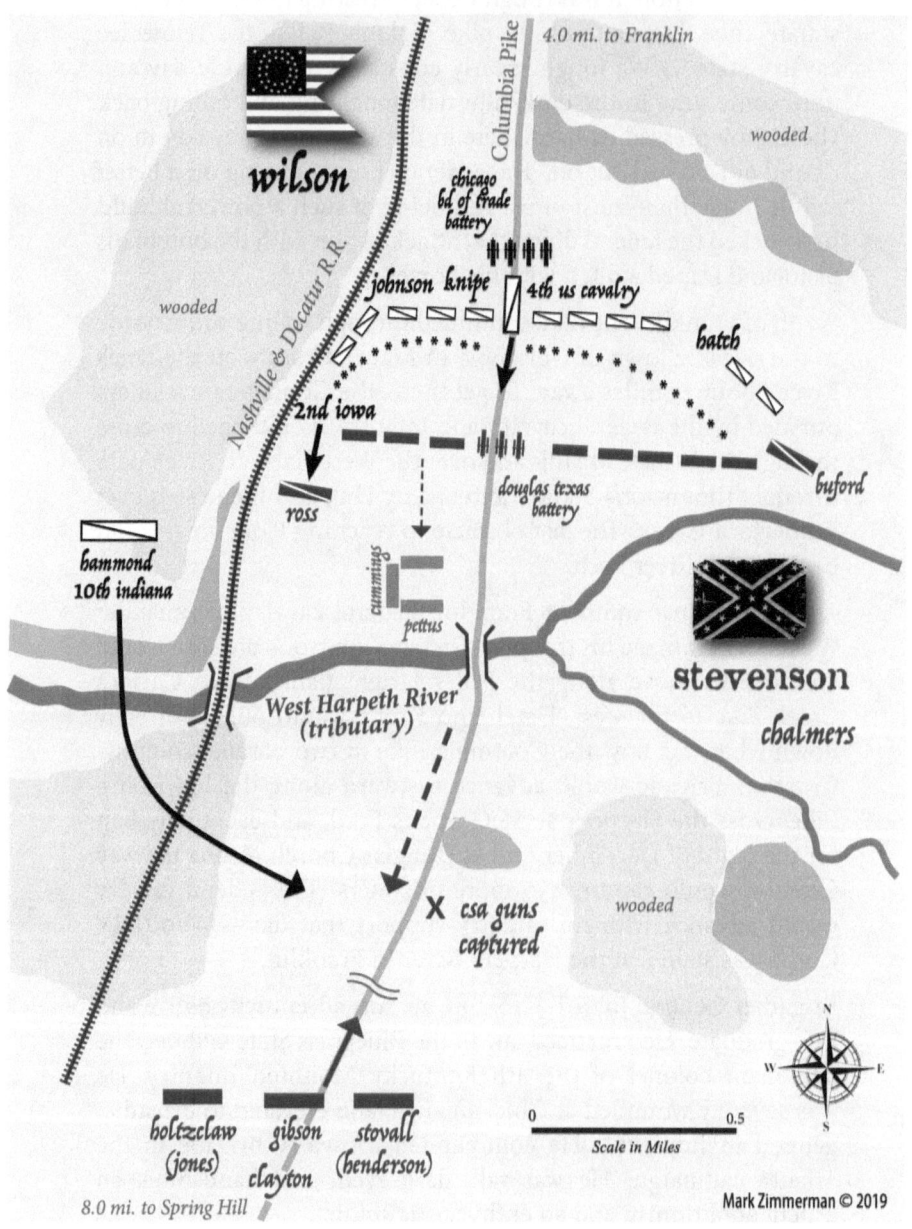

supervised the formation of the rearguard of Clayton's (Holtzclaw) men on the pike at Winstead Hill, about two miles south of Franklin. As the Federal troopers advanced, the Confederates fired three or four volleys into them and checked their advance. However, fragments from a bursting Federal shell ripped into S.D. Lee's boot, causing a painful wound. "As General Lee sat facing the enemy, I heard the ball strike the parlon of his foot," said Colonel Robert Lindsay, a regimental commander in Gibson's brigade. Then, a surgeon in their party, Dr. Stewart, was wounded. Lindsay said to Gibson, "The next two are for you and me." He was proved wrong and they retreated unharmed. Lee would remain in command for a couple more hours until he relinquished control to Major General Carter Stevenson, the senior division commander who would also direct Chalmers' cavalry.

Carter Littlepage Stevenson Jr., 47, came from a prominent Virginia family and had graduated from West Point in 1838. He saw distinguished service on the Indian frontier, in two Seminole Indian wars, the Mexican War, and the Utah (Mormon) War. During the Civil War, he quickly rose through the ranks, and was named a major general in October 1862. After the failed Kentucky campaign, he was sent to Vicksburg to reinforce Pemberton. His men fought several delaying actions and were surrendered in July 1863. Paroled, he fought again at Lookout Mountain, and several battles around Atlanta, temporarily leading Hood's corps when Hood was named army commander.

## Battle at the West Harpeth

The Federal cavalry pushed boldly through Franklin, constantly harassing the rearguard until they met Clayton's men at Winstead Hill. Then, Confederates in Stevenson's division halted about three miles south of the village and made a stand at the West Harpeth River. (The stream at this battle and on Federal battle maps is actually a tributary of the West Harpeth, which is crossed by the pike 2.7 miles farther south, where the battle ended. A modern historical sign marks the site of the battle at the river crossing just north of modern I-840. The battle likely took place 2.5 miles to the north at the tributary crossing, moved south, and ended where the modern marker stands.)

S.D. Lee noted: "Some four or five hours were gained by checking the enemy one mile-and-a-half south of Franklin (at Winstead Hill) and by the destruction of the trestle bridge over the Harpeth... About 4:00 pm the enemy, having crossed a considerable force, commenced a bold and vigorous attack, charging with his cavalry on our flanks and pushing forward his lines in our front."

Now as leader of the rearguard, Carter Stevenson ordered his infantry to make a stand just north of the West Harpeth tributary, about a mile-and-a-half south of Winstead Hill. There, the pike crosses the railroad and then rises and gently bends eastward around the nose of a low ridge. A defensive line was posted astride the pike on the reverse slope of the ridge, where the Confederates could concentrate their fire on the van of the pursuing column as it came into view. Darkness was falling, as was cold rain. Stevenson posted about 700 infantry on both sides of the pike. Positioned on the pike were the three remaining guns of Capt. James P. Douglas' Texas battery (Lt. Ben Hardin) of Courtney's Battalion. The cannon were loaded with double canister, an anti-personnel munition. The artillery charge consisted of small iron balls which spray outward from the muzzle of the cannon, converting the artillery piece into a large, deadly shotgun. Chalmers' cavalry, mostly Buford's men, were positioned on the flanks. The elevated Nashville & Decatur Railroad tracks ran roughly parallel to the pike a couple hundred yards to the west.

Down the turnpike came the Federal troopers, Knipe attacking in front, while Hatch worked the Confederate right flank and Johnson the left flank. Hatch was forced to delay his attack as some Federals got mixed in with Confederate stragglers. The Confederates moved their artillery into position during this delay. Wilson then ordered Hatch and Knipe to form ranks and charge. He directed his escort of 180 men, the 4th U.S. Cavalry commanded by Lt. Joseph Hedges, to form in columns of fours and charge straight down the pike.

Hedges moved off the pike to allow the Chicago Board of Trade artillery battery to return fire. Then Hedges crashed his 4th U.S. Regulars into the rebel line, sabers drawn and slashing. Douglas' battery opened with deadly canister, throwing the Federals backwards and leaving their commander stranded. Hedges waved his hat and yelled, "The Yankees are coming. Run for your lives!"

The ruse worked and Hedges escaped. Hedges would be awarded the Congressional Medal of Honor for his bravery. His citation reads: "At the head of his regiment (he) charged a field battery with strong infantry supports, broke the enemy's line, and, with other mounted troops, captured three guns and many prisoners." Also participating and earning a Medal of Honor was Lt. Eugene Beaumont of the 4th U.S. Cavalry, who would earn a second MOH four months later at Selma, Alabama.

Confederate Gen. Abe Buford was assaulted by a Federal trooper who slashed his saber at him. Chalmers rode to the rescue and dispatched the swordsman with two shots from his revolver. Another Federal swung at Buford, but the blade was diverted by a rebel trooper using the barrel of his carbine. Buford then grabbed the assailant, swung him onto his horse, and galloped to safety. The Federal later claimed Buford had "hugged him like a bear."

The 2nd Iowa cavalry regiment, commanded by Major C.C. Horton, was on the Federal extreme right flank near the railroad tracks. The regiment moved at a walk for 300 yards, then began to trot, and finally the charge was sounded. All sprang forward, but the center could not push through due to the steep and rocky hillside. They halted to dismount and engaged the enemy. The rest of the regiment pressed onward through a thick woods until they reached the rear of the Confederate line. The Confederate battery opened on them with grapeshot and canister, prompting Horton to call for a wheeling to the left and then a direct assault on the guns.

Horton elaborated: "(Our) horses were poor and so much blown that they could only raise a slow trot, perceiving which the enemy charged us in turn, but were handsomely repulsed with the carbine. A strong column of rebels were now reported passing through the gap between my regiment and the balance of the brigade. The fact that the day was dark and rainy, and that they wore rubber ponchos, and were many of them dressed in blue, had led my men to believe them to be our own troops, so that they were nearly in the rear of the 3d Battalion before the mistake was discovered."

Apparently, Horton's mounts had reached their limit of endurance, but the troopers' firepower, implemented through the use of their Spencer repeating rifles, saved them from disaster.

Bronze sculpture (1924) depicting a Federal cavalry charge, by Henry Shrady, in front of U.S. Capitol.

One company of the 2nd Iowa regiment, commanded by Sgt. John Coulter, was nearly surrounded by the cavalry of Lawrence Sullivan "Sul" Ross, a brigade in Red Jackson's division, and forced to cut and slash their way out with the saber. The sergeant, along with a corporal and two privates, charged the rebel color guard in their front and captured the colors of Ross' brigade after a desperate hand-to-hand struggle. The corporal and one of the privates were killed; Coulter and the other private were wounded. The regiment lost seven killed, eight wounded, and 13 captured. The regiment captured 50 prisoners. The 2nd Iowa lost a color bearer, the second one in as many days.

Horton added, "Eight dead rebels, lying within the space of a few yards, attest the desperate nature of the conflict. After a few moments of close fighting, in which the sabre and butts of guns were freely used, the rebels fell back."

Stevenson claimed that it was impossible to control his own cavalry, which retreated in disorder, "leaving my small command to their fate." He added, "This was a critical moment, and I felt great anxiety as to its effects upon the men, who, few in numbers, had just had the shameful example of the cavalry added to the terrible trial of the day before."

With pressure on his left flank and rear, Stevenson formed his men west of the pike into a three-sided square (a hollow square is the traditional infantry tactic to repulse cavalry). Colonel Watkins, commanding Cumming's brigade, refused his line against the flank attack while General Pettus joined at right angles, facing southward. Hammond's Federal brigade, with Colonel Benjamin Gresham's 10th Indiana in the lead, forded the West Harpeth tributary upstream and hit Stevenson in the flank.

Stevenson was forced back and crossed the stream, fighting every step of the way. Finally, with near darkness at hand, he briefly halted to reorganize his lines, only to be struck again by Hatch. In the darkness, men could not discern friend or foe. Units mingled. Men were wearing slickers, and some Confederates wore blue jackets. For fierceness, the skirmishing exceeded any his regiment had ever engaged in, wrote a veteran sergeant of the 2nd Iowa cavalry.

Farther down the road, Knipe's troopers forced their way

across the stream. The pressure on the rearguard was too great. Stevenson's men hustled down the Columbia Pike, leaving behind the three 12-pounder guns of the Douglas Battery. To the south, Clayton, fallen back from Winstead Hill, hearing the fight, formed his men across the road and opened fire, virtually in the faces of Hammond's men. The Federals fell back in disorder, leaving a stand of colors. Col. George W. Jackson, commander of the 9th Indiana Cavalry, Hammond's Brigade, was injured by a falling horse during this action.

When Holtzclaw's brigade advanced to help the remnants of Stevenson's command withdraw, they collided with the 9th Illinois cavalry of Hatch's division. Holtzclaw's men retreated to the vicinity of Thompson's Station and bivouacked for the night with the rest of Lee's troops.

Clayton wrote in his after-action report: "When in about 100 yards of the left of General Gibson's command, which rested upon the pike, I saw a column of cavalry moving obliquely and just entering the road a few paces in my front. An infantry soldier of my command, recognizing me (it being then quite dark), ran up to me and whispered, 'They are Yankees.' Turning my horse to the left, so as to avoid them, I moved rapidly to the right of General Gibson's line, and after narrowly escaping being killed by several shots fired at me through mistake, I communicated the information to General Gibson, who promptly wheeled his brigade to the left and delivered a volley which scattered the enemy, killing many of them. I then, at the suggestion of General Gibson, moved back these two brigades behind a fence, in order to better resist a charge and also for greater security against firing into our own men. This position was scarcely taken when the enemy again began to move from the left upon the pike in our immediate front. Demanding to know who they were, I was promptly answered, 'Federal troops,' which was replied to by a volley, killing several and again driving them off, leaving a stand of colors, which was secured. The enemy having finally retired and the firing having ceased..."

Wilson reported: "We have 'bust up' Stevenson's division of infantry, a brigade of cavalry, and taken three guns. The Fourth Cavalry and Hatch's division, supported by Knipe, made several beautiful charges, breaking the rebel infantry in all directions. There has been a great deal of night firing, volleys and cannonading from our guns — the rebels have none. It is very dark, and our men are considerably scattered, but I'll collect them on this bank of the stream — West Harpeth."

Exhilarated but winded, Wilson was forced to call off the pursuit due to exhaustion and darkness. He exclaimed, "Hatch is a brick! If it had only been light we would certainly have destroyed their entire rear guard; as it was, they were severely punished."

Wilson said the West Harpeth affair was "another running night fight, in which all semblance of order was lost, where regiment got separated from regiment, troop from troop, and officers from men. There was no guide but the turnpike, and no rule but 'when you hear a voice shoot, or see a head hit it.'"

In his book on Civil War retreats, historian David Frey stated, "It was by almost all accounts a tremendous victory for Federal cavalry, but darkness helped frustrate the fulfillment of Wilson's primary objective—the capture of Hood's rear guard."

By this time, after dusk on Saturday, December 17th, the main body of Hood's army, ragged and exhausted, had reached Spring Hill, on the Maury County line, to spend the night. They had marched 21 miles that day from Hollow Tree Gap. The soldiers could not take any route other than the congested turnpike due to the mud and muck. One can only try to imagine the conditions following the passage of hundreds if not thousands of horses and mules. Back in Franklin, Wood's IV Corps sat on the north bank of the Harpeth River waiting for the hardluck pontoon train to arrive.

Private Stephenson of the Washington Artillery complained about the persistent cold rain: "The rain still poured in torrents upon us, more dogged in its pitiless pursuit than the enemy. It still beat us down, as it had been doing day and night, day and night, ever since the day of our defeat, until the drops felt like heavy shot upon our heads." He noticed that Hood's army was slowly gathering its wits. "Gradually therefore, as the army found itself unmolested, attempts at order were made, the broken ranks straightened, commands separated to themselves, and a semblance of discipline arose." Pvt. Stephenson arrived in Spring Hill at 10:00 pm and slept in a barn on the estate of Ferguson Hall, the home where Confederate General Earl Van Dorn had been shot and killed by a jealous husband in 1863.

Hood took note of the day's whirlwind of events: "During this day's march the enemy's cavalry pressed with great boldness and activity, charging our infantry repeatedly with the sabre, and at times penetrating our lines."

"Hood's main army was thoroughly whipped and dispirited," noted historian Horn, "but the rear guard during those trying days was indeed undaunted and firm, and deserving of every possible word of praise for a prodigious performance under the greatest difficulties."

Meanwhile, back in the nation's capital, U.S. Grant was not satisfied, and he let General Thomas know his feelings. The crusty Thomas replied, with all due respect but blunt sincerity, "I am doing all in my power to crush Hood's army, and, if it be possible, will destroy it, but pursuing an enemy through an exhausted country, over mud roads, completely sogged with heavy rains, is no child's play." Indeed.

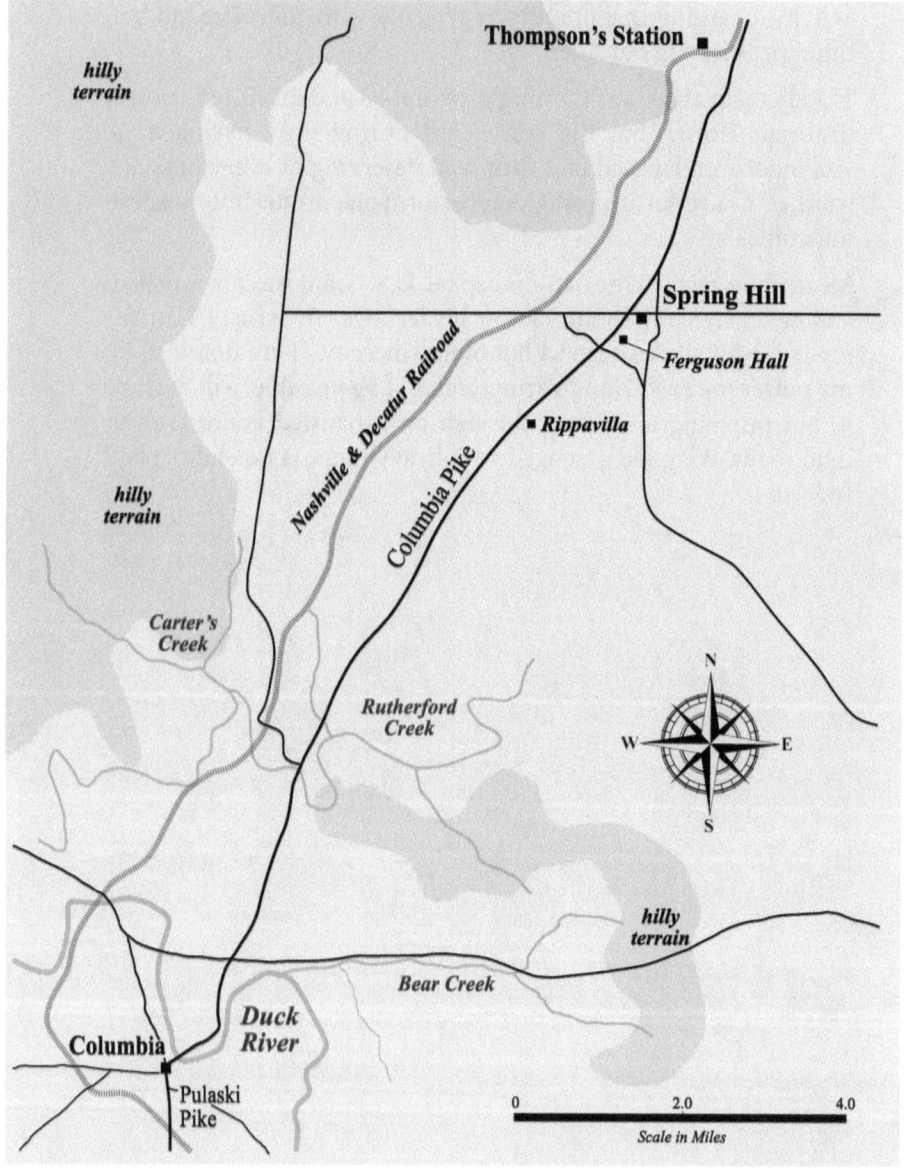

## Forrest Rejoins Hood
*Sunday, December 18th, 1864*

After three days of more or less continuous battle, both the Federals and the Confederates were beginning to feel the strain. Yet onward they trudged. Hood's army would be reaching the relatively large town of Columbia on Sunday the 18th, and although skirmishing continued unabated, there would be no major confrontations that day.

At or about sunrise on Sunday, the sore, tired, and soaked Southern soldiers and troopers scattered along a broad swath of turnpike around Spring Hill awoke to find the rain had stopped, at least temporarily. Then, as if to torment them, the downpour resumed at 7:30 am and did not quit until 3:00 pm. Hood's men shook themselves awake and contemplated marching a dozen miles to the Duck River and crossing over into the town of Columbia on the south bank. There might be campfires and even hearths there, with soft beds and tables groaning with fresh food and tasty whiskey. Thus, some daydreamed of their destination only to be startled back to reality and wonder how far behind were the pursuing blueclad troopers. There was, however, one obstacle in their path to the river—the inundated, swollen waters of Rutherford Creek. Despite the dangers, most of the Confederates managed to cross the rapid waters on makeshift or pontoon bridges.

The Federal pursuit resumed early on the Sabbath, men and beasts arising from some well-deserved and much-needed rest. Richard Johnson's men turned east from the Carter Creek Pike to Spring Hill and hit the Confederate rearguard in flank, but his cavalry force was not strong enough to hold the gray troopers at bay. By 11:00 am, however, the Federal cavalry had chased all of the Confederate stragglers from their Spring Hill bivouac.

Early in the morning, Wood's infantry corps finally began crossing the Harpeth River at Franklin, a makeshift bridge having been hastily built by the veterans of the 9th Indiana infantry. Wood's men marched 18 miles that day, not reaching the enemy but advancing a mile ahead of the Federal cavalry, and at 4:45 pm the corps bivouacked about 3.5 miles north of Rutherford Creek. The creek was 15 feet deep at most places, raging, and unfordable. Men would need to cross to the opposite bank in order to begin building a bridge. Meanwhile, the bluecoat infantry corps of Major Generals A.J. Smith and John Schofield lingered at Franklin.

After three consecutive days of battle, with the Confederates now positioned between Rutherford Creek and Duck River, the question still lingered—where the heckfire was cavalry chief Major General Nathan Bedford Forrest? Rumors had him most everywhere, lurking around the next hill and ready to strike. On December 17th, both Schofield and General Rousseau at Murfreesboro reported to Thomas that the "wizard of the saddle" had been killed. Harry Wilson, however, the man whom Forrest had bested on the march north and the man who had believed Rucker's assertions at the barricade, assured the Federal commander that Forrest was very much alive and en route to Columbia from Murfreesboro.

On December 18th, the Mississippi brigade of Confederate troopers led by Brigadier General Frank Armstrong reinforced Abe Buford by reaching Spring Hill. Armstrong was known as a competent commander and was famous for having led a company of Federal dragoons at First Manassas before deciding to switch sides and join the Confederacy. Under Van Dorn, he was instrumental in the Confederate victory at Thompson's Station in early 1863.

Forrest and his remaining small command (including Smith's infantry brigade led by Colonel Charles H. Olmstead, Colonel J.B. Palmer's brigade, and several artillery batteries) moved ten miles west to Triune, where his wagon trains, wounded, captured Federals, and cattle waited. This assorted assemblage wound south and tried to ford the Duck River at Lillard's Mill (about six miles west of Forrest's birthplace at Chapel Hill). On December 18th, Forrest and a small contingent managed to cross the raging waters and arrived at Columbia that night. Forrest then joined Chalmers, who was holding the line at Rutherford Creek, four miles north of Columbia. Most of Forrest's men crossed the river back into Columbia that evening. Cheatham's infantry crossed the creek and then formed a defensive line, allowing Tyree Bell and his troopers to ford the creek a distance from the pontoon bridge. Once across Rutherford Creek, Bell's men were out of the fire of the enemy for the first time in three days and nights. Major Tom Allison appeared with provisions that he and his men had collected from his brother's home near Franklin. "We had cracklings, back-bone, spare ribs, etc., everything good at hog-killing time," Bell noted with glee.

Columbia (pop. 4,000) was the seat of Maury County, the richest antebellum county in the state based on its cotton plantations and agricultural wealth. Later, it would become known as the mule capital of the world. James K. Polk practiced law at its town square courthouse before becoming U.S. President in 1845. Sitting on the south bank of the Duck River, Columbia was a major stop on the turnpike and the railroad line. Many fine mansions festooned with Greek columns adorned the town.

As the majority of Hood's troops concentrated at Columbia, Cheatham's Corps remained in line of battle on the south side of Rutherford Creek after destroying or removing all the bridges. Cheatham and Chalmers held the creek line all that day. A.P. Stewart arrived at Duck River on the morning of December 18th and formed on the north bank, Major General William W. Loring on the right, and Walthall on the left, covering the passage of the army. The Confederate soldiers filing into town were not met with any great enthusiasm by the local civilians. One resident of Columbia wrote in his diary: "They are the worst looking, and most broken down looking set I ever laid eyes on."

At 1:00 pm on December 18th, a significant development occurred. Federal cavalry chief Wilson called a halt to the pursuit seven miles north of Columbia, three miles north of Rutherford Creek, and would not resume operations in full measure for several days. The Federal cavalry went into camp, the spring finally wound out of its coil. Why would an overly ambitious officer such as Wilson, so eager to please and gain glory, basically lay down? Was this not the vicinity where Forrest had bested him three weeks earlier? Wilson knew that Hood's men were crossing the Duck River on pontoon bridges. The Federal pontoon bridge had yet to be delivered, time of arrival unknown. Wilson's troopers had been fighting now for basically three straight days; they were flat-out dog-tired and needed a rest. Supplies and ammunition needed to be replenished. As stated previously, a retreating army falls back on its own supply line, while a pursuing force tends to outrun its own supplies. But the main reason seemed to be that the cavalry horses could not take much more abuse without collapsing. The hilly terrain, the mud and the slop, the skirmishes and ambushes, and the need for sustenance in the miserable wet winter weather proved to be overwhelming. There is much evidence to indicate that the Federal horse stock was not of the highest quality in the first place—more coursers and rounceys than stallions or destriers. Horses needed to be tended, groomed, and reshod. Tack and gear needed to be repaired; weapons cleaned.

Cavalry advancement with an alternating walk (four miles per hour), then trot (eight mph), can cover 35 miles a day under ideal conditions. This pace can be sustained for five days before a day of rest is required. The December winter weather, muddy terrain, and short days were not ideal conditions, however, to say the least.

Horses require a lot of feed. "Horses are eating machines," noted equine expert Lt. Col. (ret.) Edwin Kennedy. Quartermaster regulations called for 26 pounds of food (14 pounds of hay and 12 pounds of grain, usually oats, corn or barley) per day per horse (23 per mule). That would total 360,000 pounds per day for Wilson's horses. One supply wagon ordinarily carried one ton or 2,000 pounds, so roughly 180 wagons would be needed by Wilson's horses per day. The horses pulling the forage wagons would require an additional nine tons of forage per day. A single horse could pull 1,900 pounds 20 to 23 miles a day over a macadamized road, such as the Columbia Pike, but only 1,100 pounds over rough ground (due to the rainy weather and mud, it was impossible for wagons to move anywhere off the turnpike). It's difficult to imagine how Wilson's cavalry and Wood's infantry could be adequately supplied on a daily basis by wagon trains all using the same main road. Of course, the railway was used to supply the troops, but its usefulness is not adequately known.

Federal cavalrymen were supplied horses; Confederates usually used their own horses (and were supposedly paid 45 cents per day for feed). This led Federal troopers to abuse their mounts more than a Southerner might. On the other hand, knowing the difficulty in obtaining a remount, a Confederate might be less willing to lead his horse into harm's way.

Before crossing the Duck River, the Confederate rearguard withstood two attacks that day that were easily beaten back. As Cheatham's Corps prepared to cross the pontoon bridge, Forrest and his men arrived, with the same intention. Forrest asserted his right to cross first. Cheatham replied, "I think not, sir. You are mistaken. I intend to cross now, and will thank you to move out of the way of my troops." Forrest drew his pistol and rode over to Cheatham and said, "If you are a better man than I am, General Cheatham, your troops can cross ahead of mine." Soldiers in both commands raised their weapons to defend their commanders. Cheatham reportedly replied, "Shoot! I am not afraid of any man in the Confederacy!" General Lee stepped in to mediate, and convinced the two chieftains to apologize. It is not known for certain which group crossed first although one report asserts that Lee sent Forrest across first and then pacified Cheatham.

Forrest was not a man to be challenged. "War means fighting, and fighting means killing," he once said. Earlier in the war, in June 1863 at Columbia, Forrest tangled with a subordinate, Lt. Andrew Wills Gould, who accused Forrest of calling him a coward for losing several cannon in recent combat. Gould shot Forrest, who thought he was mortally wounded. Shouting an oath that no man could kill him and survive, Forrest drew a folding knife, opened it in his teeth, and plunged it into Gould's chest. Long story short, Forrest survived, but Gould died of his wound.

Back at Hood's headquarters at the Vaught mansion in Columbia, Chaplain Charles Quintard of the 1st Tennessee said that he and Lieutenant Colonel Johnston advised Hood that the Army of Tennessee should hold the line at the Duck River, if possible. Quintard argued that to fall back across the Tennessee River in Alabama would dispirit the men, cause mass desertions among the Tennessee troops, and generate enthusiasm in the North to prolong the war, "whereas if we can hold this line, put the machinery of our state government in operation, the campaign, even with our reverses, will be a splendid success."

Major James D. Porter of Cheatham's Corps disputed Yankee claims about the Army of Tennessee: "The successful resistance to the assault of the Federal cavalry near Franklin by the rear guard of

Lee's Corps, repeated the next day by the rear guard of Cheatham's Corps, does not sustain the Federal general's (Thomas) report that our army was a disorganized rabble."

But Hood was not yet sure that he wanted to stay put at Columbia and make a stand there. At this point, the Tennessee River was still 70 miles away.

Southerners in the main body of the army may have been miserable but they probably also felt fortunate, recalling the fate of some of their comrades and their travails to come.

Pvt. Samuel Boyd of the Charlton Rebels, 45th Mississippi Infantry, was wounded December 15th at Nashville and captured three days later at Franklin. He was sent to Hospital No. 1 in Nashville and then to Camp Chase Prison in Columbus, Ohio. He died March 2, 1865 of pneumonia and is buried in the Confederate Cemetery there.

Pvt. William B. Tomlinson of the 17th Alabama Infantry was captured Dec. 17th at Franklin and sent to Camp Chase, where he died of pneumonia in March 1865. Pvt. William A. Beck of the 36th Georgia Infantry died at Camp Chase of variola (smallpox). Pvt. Levi Victory, aka Victor, of the Harper Guards, 42nd Georgia Infantry, died in prison of chronic diarrhea. Pvt. Joseph Babin of the 4th Louisiana Infantry died in captivity Feb. 7, 1865, cause unlisted. Pvt. William J. Poindexter of the 20th Tennessee Cavalry died in March 1865 of pneumonia. All are buried in the Confederate Cemetery at Columbus.

Other captured Confederates were luckier and survived the camps. They were released in the spring or summer of 1865 after taking the oath of allegiance.

## Stymied By Rising Waters
*Monday, December 19th, 1864*

The weather in Middle Tennessee, even in December, can change quickly and drastically with an alteration of the prevailing winds. On December 19th, the southwest winds which had brought mild weather began to emanate from the northwest, adding a frosty chill to the intermittent rain. The temperatures dropped steadily that day and by nightfall the landscape turned intensely cold. The drop in the temperatures would bring only added miseries to the fleeing Confederates and enhance the obstacles faced by the pursuing Federals.

While the Federals pursued Hood on land, they also attempted a broad flanking movement, by water, 70 miles south of Columbia. On the evening of December 18th, General Thomas wired Acting Rear Admiral Samuel P. Lee to request that Federal gunboats on the Tennessee River destroy the bridges at Florence, Alabama, Hood's supposed escape route. At that point, the river could not be forded or crossed in any way other than a pontoon bridge or river ferry.

Meanwhile, the Federal IV Corps led by T.J. Wood moved in advance of the Federal cavalry and reached troublesome Rutherford Creek at 9:30 am. The creek was reported to be 15 feet deep at most places with a strong current. Lieutenant Colonel Joseph S. Fullerton summed up the situation: "The pontoon train has not yet come up and we can hear nothing of it. We have not the tools to build a bridge what wagons can cross on. The rain still continues to fall very fast and the creek is yet rising rapidly...the rain has ceased now, and it is blowing up quite cold."

Wood spent a miserable day at Rutherford Creek: "Rafts were constructed and launched, but the current was so rapid that they were unmanageable. Huge forest trees growing near the margin of the stream were felled athwart the stream, with the hope of spanning it in this way and getting some riflemen over; but the creek was so rapid and the flood so deep that these huge torsos of the forest were swept away by the resistless torrent. In these efforts was passed one of the most dreary, uncomfortable, and inclement days I remember to have passed in the course of nineteen-and-a-half years of active field service."

At least one day was lost at Rutherford Creek due to the misdirection of the pontoon train. During the night of December 19-20th, Federal engineers somehow managed to construct two floating bridges for the infantry to cross. The infantry crossed and marched to the north bank of Duck River three miles away. The Federal pontoon train finally caught up to the infantry between midnight and daylight.

During this time, the Federal XVI Infantry Corps of Andrew Jackson Smith moved southwards from Franklin with its trains, leaving Schofield behind in the village.

Meanwhile, Wilson was stuck. All of the Federal cavalry except Hatch was out of rations and ammunition. Wilson's supply wagons had trouble moving on the pike due to the profusion of infantry wagons. Although Thomas ordered the cavalry to remain in bivouac that day, Hatch was directed by Wilson to venture onward. Without the pontoon train, Rutherford Creek, now more like a river, remained a formidable obstacle to the pursuit. The current was swift, the banks were steep, and the creek bed impossible to ford. All day long, Cheatham's men on the south bank fired at Hatch's troopers and Wood's infantry.

Hatch's troopers scoured the countryside, searching for a suitable place to cross the creek. The Federal cavalry was stymied by the various secondary streams and branches in the area. Throughout the afternoon, Hatch's men worked in the rain, piecing together logs and timbers from barns and outhouses, and felling trees. Once across their makeshift bridge, they discovered they had crossed Carter's Creek and were still on the wrong side of Rutherford Creek. Several men who tried to cross on a raft failed and drowned. Hatch ordered his men to dismount and move upstream one mile. Coon was able to cross a handful of the 6th Illinois on remnants of the burned-out railroad bridge, but he could make no crossing in force. Coon found a ford farther upstream, but it turned out to be

Curtis Creek instead of Rutherford Creek.

Late on December 19th, Wilson finally succeeded in getting some cavalry across Rutherford Creek. Cheatham's men fell back, skirmishing, and crossed over the Duck River into Columbia.

Forrest had reached Hood's headquarters at the Vaught house in Columbia before sunrise on December 19th. "His opinion is that General Hood ought to withdraw without delay south of the Tennessee," said Chaplain Quintard. "That if we are unable to hold the state, we should at once evacuate it...General Hood decided to fall south of the Tennessee (River). Governor Harris, in whose judgment I have great confidence, thinks it is the best we can do."

Hood "expressed the belief that he could not escape in such weather, with unfavorable roads and broken-down teams." Forrest said that "to remain (in Columbia) would certainly result in the capture of the whole force," but that, if reinforced with 4,000 infantry, he would undertake to secure time and opportunity for the escape of the army across the Tennessee. Forrest stated, "If we are unable to hold the state, we should at once evacuate it." Exiled Governor Isham Harris agreed.

Harris, 46, an attorney and longtime politician who had been governor of Tennessee when the state seceded (he was a major proponent of secession), now traveled with the Army of Tennessee as an aide-de-camp to Hood, the capital having been captured by the Federals. Harris had been at the side of General Albert Sidney Johnston when he died of wounds at Shiloh.

At Columbia, Hood was asked by some of his soldiers about furloughs. A known gambler, Hood responded, with poker in mind, "After we cross the Tennessee. The cards have been fairly dealt, for I cut them and dealt them myself, and the Yankees have beat us in the game." A soldier of the 19th Tennessee replied, "Yes, General, but they were badly shuffled."

Hood ordered Forrest to assume command of the rearguard of the army. With Chalmers' wholehearted support, Forrest asked that the rearguard infantry be led by Major General Edward Walthall, and Hood agreed. At 33, Walthall, a Mississippi attorney before the war, was the youngest major general in the Western Theater. Chalmers cited Walthall's knowledge of tactics, discipline, his care and attention to his men, talent and ability to command, and courage in the face of the enemy. Walthall had supported Forrest in Winfrey Field at bloody Chickamauga, employing "aggressive action against overwhelming odds." Walthall also had skillfully helped in covering the Confederate retreat at Missionary Ridge. In February 1864, Forrest had offered Walthall command of a cavalry division, but Walthall declined, citing his lack of cavalry experience.

The Confederate troops were in bad shape — scantily clad, poorly armed, and tired and dispirited. One Tennessee soldier was overheard, grousing, "Ain't we in a hell of a fix: a one-eyed President, a one-legged general, and a one-horse confederacy!"

"Clearly the Confederates failed to meet basic nutritional needs for man and animal alike," according to historian Kimmerly. Critics have attested that Hood never comprehended the complexity of coordinating logistical support for his troops in Middle Tennessee. To be fair, materials were in short supply by that time."

The arrival of Forrest at Columbia and his subsequent leadership of the rearguard had a significant impact on the mood and spirit of the retreating Confederates. In an address in 1879, Chalmers noted that Forrest was well aware of the danger the Army of Tennessee faced, but he had assumed responsibility with vim and vigor. "Such a spirit is sympathetic," said Chalmers, "and not a man was brought in contact with him who did not feel strengthened and invigorated, as if he had heard of a reinforcement coming to our relief."

Having fought and slashed their way 38 miles from Compton's Hill to Columbia, Hood's army was now a little over 70 miles from the Tennessee River crossing at Bainbridge, Alabama, a tiny village and ferry six miles east of Florence. All of the pontoon-boats in Hood's wagon train would be needed to build the lengthy floating bridge. The question was — would those be enough?

Hood established his headquarters that night at Major A.W. Warfield's imposing Greek Revival mansion three miles south of Columbia on the pike. The home is now known as Beechlawn, for the enormous beech trees that surround it. Hood had quartered there on the night of November 27th during the northern advance.

During the night of December 19-20th, A.P. Stewart's Corps completed a harrowing crossing of the Duck River, which rivaled

Washington's crossing of the Delaware, according to Gale. At daylight the next morning, the Confederate engineers removed the pontoon bridge. All of Hood's army was now safely over the Duck River.

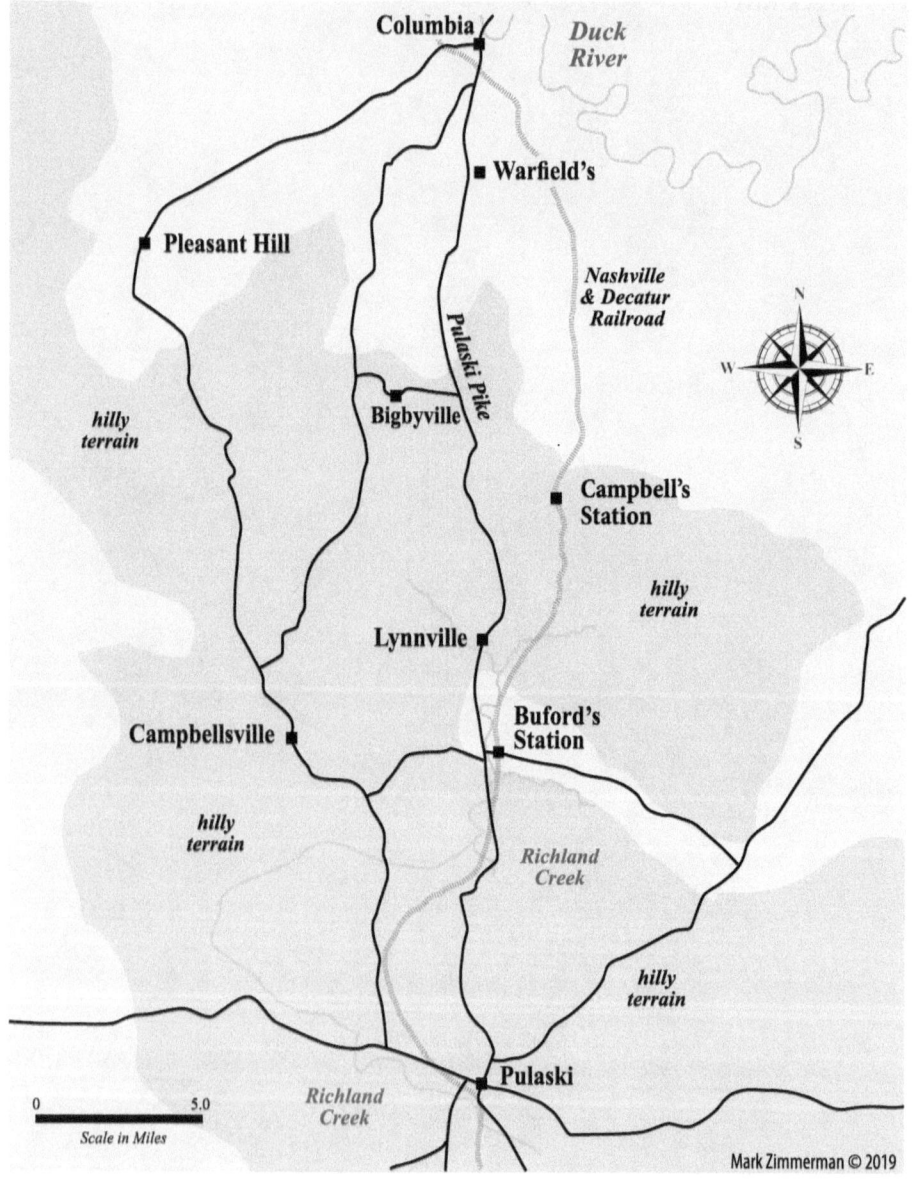

## The Rearguard Reorganizes
*Tuesday, December 20th, 1864*

By the morning of Tuesday, December 20th, the sleet had turned to snow as Hood and the main portion of his army left Columbia, having put Forrest in charge of the rearguard with orders to hold the Duck River line as long as possible. The next destination down the turnpike was the small town of Pulaski, 28 miles to the south, where both the paved turnpike and the railroad came to an end. South of Columbia, the soldiers found the rolling countryside turning into a much more hilly and barren terrain. The narrow defiles through the hills would restrict movement along the flanks and provide ample opportunities for setting up an ambush.

The march of the Army of Tennessee, miles in length, was resumed on the Pulaski Pike with S.D. Lee's Corps in front, commanded by Stevenson; Cheatham's Corps next; and A.P. Stewart's Corps bringing up the rear. From the perspective of a hawk soaring high above, the main road must have appeared to be a gigantic moving thing, a serpent rippling forward in fits and starts. By the end of the frigid day, the head of the column camped within two miles of Pulaski with the three corps bivouacking in order of their march.

On the morning of the 20th, Hood summoned the young yet experienced Walthall to command the infantry portion of the rearguard. Walthall noted, "I have never asked for a hard place for glory nor a soft place for comfort, but take my chances as they come. I will do my best." Hood replied, "Forrest wants you, and I want you."

Walthall then organized his troops. His rearguard infantry consisted of eight brigades, nominally more than 30 regiments— four of Cheatham's brigades, two of his own (Reynolds and Quarles), Ector's Texans under Coleman, and one of Loring's—

Featherston's Mississippians. The brigades, totaling 1,920 riflemen, were consolidated as follows:

- Featherston and Johnston under Brig. Gen. Winfield S. Featherston (498 men);
- Coleman and Reynolds under Brig. Gen. Daniel H. Reynolds (528 men);
- Heiskell and Feild under Colonel H.R. Feild (278 men); and
- Olmstead and Palmer under Colonel J.B. Palmer (616 men).

Forrest had 1,600 effective infantry (not counting the 400 men marching without shoes) and 3,000 cavalry to face a pursuing Federal force of more than 10,000 cavalry armed with repeating rifles and perhaps up to 30,000 infantry.

The 400 men without shoes? One of Walthall's staff officers reported: "The sufferings of the troops were terrible...Without protection from the severity of the weather, without blankets, and many without shoes, and nearly all indifferently shod, the horrors of the retreat were to be seen as the bare and frostbitten feet of the soldiers, swollen, bruised, and bloody, toiled painfully over the frozen pike."

A master tactician, Forrest devised a workable solution. At Columbia, Forrest left half the wagons along the pike and doubled the teams for the other half. Then the teams returned and hauled the rest of the wagons away. He did this during the two days the Federals needed to cross Rutherford Creek and the Duck River. Then Forrest used the wagons to haul the shoeless infantrymen of the rearguard until they were needed to fight. He secured as many oxen as he could find to haul the wagons and sent the rest of the cargo wagons off with the main body of the army.

However much the hardships and privations, these men were rugged and hardened veterans. "The usage and customs of war, and its privations had inured them to such hardships as but few men could bear, and made them veteran soldiers," wrote Ralph J. Neal in his history of Co. E, 20th Tennessee. By this point, those Confederates inclined to surrender or desert had done so already.

Artillery commander Lt. Joseph Chalaron noted that "Corporal D.A. Rice, a gunner in the Washington Artillery, had been wounded in the head at Kennesaw Ridge so that is was impossible for him to close one eye, and the cold striking it, kept it with flowing tears continually, tears that froze and formed a pendant icicle six inches long at times."

Forrest also had eight artillery pieces under Capt. John W. Morton. Morton was only 22 years old but he had served capably as Forrest's chief of artillery for several years. Morton joined the army as a cadet at the Nashville Military Institute and was captured at Fort Donelson as a member of Porter's Battery. After months in a Northern POW camp, he was exchanged and joined Forrest in the fall of 1862. Due to his youthfulness and slight build, Morton had to work twice as hard to gain Forrest's confidence.

At mid-morning on December 20th, with Cheatham's gunmen now gone, the pioneering men of Wood's IV corps and Hatch's division fashioned two makeshift footbridges over Rutherford Creek, one using the remains of the old wrecked railroad bridge. (The 58th Indiana Regiment was composed of trained pontoniers and bridge-builders, but unfortunately they were with Sherman in Georgia.) Hatch's cavalry took the lead and arrived opposite Columbia in the early afternoon, along with the 7th and 9th Illinois regiments on foot.

Wood reported: "During the night (of the 19th) and the early forenoon of the following day, the 20th, two bridges for infantry were constructed across the stream—one at the turnpike crossing, by Colonel Opdycke's brigade, of the Second Division, and the other by General Grose's brigade, of the First Division. So soon as these were completed the infantry of the corps was passed over, marched three miles, and encamped for the night on the northern bank of Duck River."

Wood's infantry reached the north bank of the Duck River at 2:00 pm, having been delayed 34 hours waiting for the pontoon train to cross the Harpeth River and Rutherford Creek.

Marching took its toll. Wood requested 15,000 pairs of shoes and socks from Nashville warehouses for his weary men. Their shoes, most likely of dubious quality in the first place, had been ruined by the rain and the tromping on the rough turnpike. The mud and muck churned up during the day froze into sharp ruts and furrows during the night, making foot travel precarious. Wood claimed that many of his men would be disabled within days without a new supply of footwear.

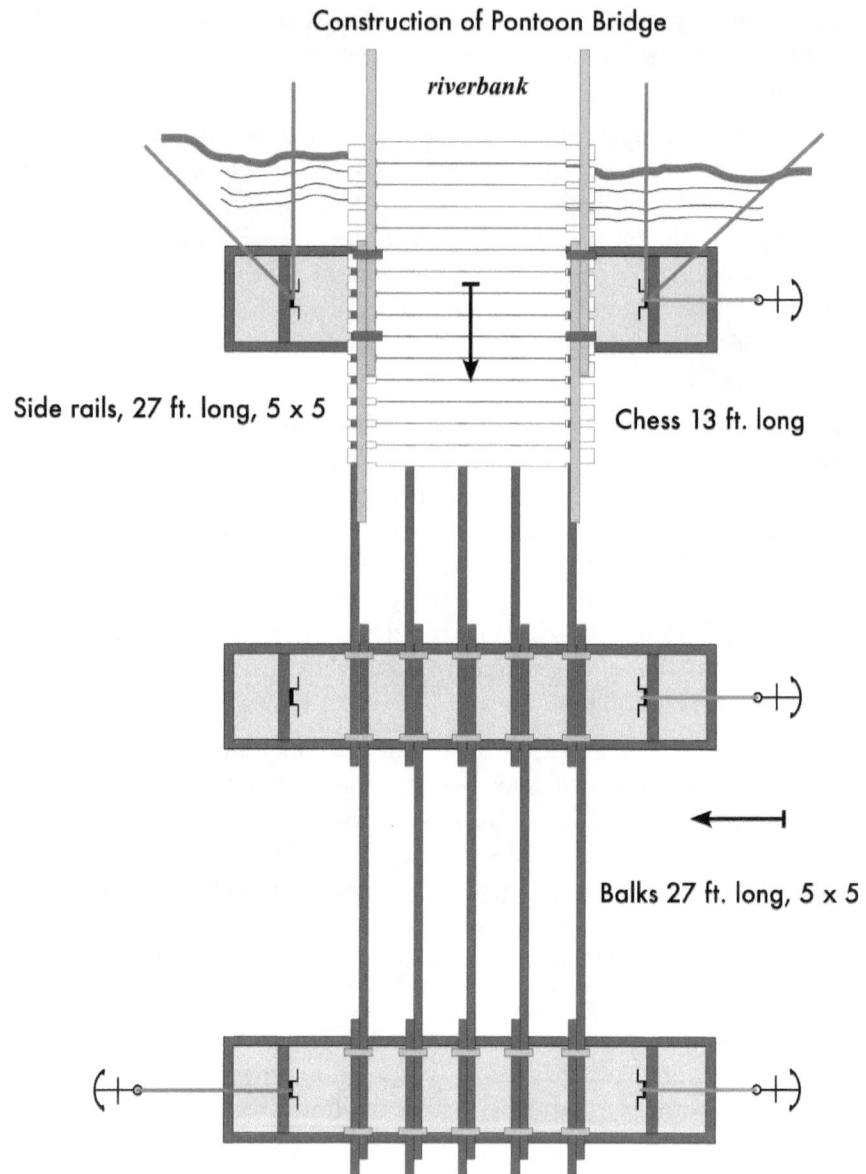

Two types of Federal wooden pontoon-boats on their carriages. (Library of Congress)

Federal cavalry leader Wilson lost two days building bridges, and three more days were needed to cross the Duck River while they waited for more rations and ammunition. Even the railroad wasn't functioning properly due to the delays in reconstructing the railway bridge at Franklin. Hammond, Croxton, and Harrison remained in camp drawing supplies. Under orders from Wilson, the remainder of Richard Johnson's and Knipe's troopers made their way back to Nashville to find new horses to mount. Captain Orlando H. Sheaver of Fielding Hurst's 6th Tennessee U.S. Cavalry said his men "pursued the enemy to near Springhill when we went back, reached Nashville on the 21st, Wet, Cold, Hungry and Pised." Wilson also sent two dismounted brigades back to Louisville for mounts.

General Thomas, now at Rutherford Creek, said that A.J. Smith's men would assist in getting the pontoon train forward—at 6:00 pm that day it was passing through Spring Hill. Many of the pontoons were basically long boats; others consisted of rafts and wagon beds covered in canvas.

Hatch stopped at the Duck River and began a furious artillery shelling of Columbia. By this time, the only occupants of the town were some of Forrest's men, the wounded of both sides, and Federal prisoners. Forrest rode to the riverbank under a flag of truce and persuaded Hatch to stop the bombardment. Forrest also offered to exchange 2,000 Federal prisoners who had been suffering from exposure to the elements (the Confederates had no blankets or clothing to spare). Within two hours, Thomas himself replied, refusing to accept the offer. Thomas said most of the rebel prisoners had already been sent north and therefore could not be exchanged. But he also probably reckoned that an exchange would favor the Confederates, that having to care for 2,000 Federal prisoners would hinder their retreat. (It is not clear what became of those prisoners.)

As much as the soldiers of both sides suffered during the retreat, worse off were the civilians who remained in the towns and villages and on their farms. In rural Tennessee, the civilians had already suffered through three years of war. In destructive power, nothing compared to the movement of a large army through the countryside, regardless of the flag flown. Nimrod Porter, a 73-year-old plantation owner near Columbia, said the Army of Tennessee stole hogs and other foodstuffs and burned his fence rails for firewood. But the Federals were worse. During Hood's retreat, "Croxton's Yankees came through and stole everything. They cooked the last old gobbler and all the chickens, over a fire. They even took the boots off the blacks. Last night they took all of black Sukey's money, all my corn and what little oats I have left ... A gray fox ran under the kitchen walk. I shot it for dinner. We have a little parched corn."

During the first half of the 19th Century, chattel slavery and white supremacy were the way of life in the South. Men and women under bondage suddenly finding themselves displaced from their masters followed the Federal army wherever it roamed and were known as contrabands. Most were put to work at menial labor, digging trenches, tending to livestock, or working as teamsters. Some were recruited or joined the ranks of the newly founded U.S. Colored Troops, which were led by white officers.

Nothing frightened white Southerners more than the threat of a violent insurrection by their black slaves. Nothing enraged Southern soldiers more than an armed black man in a blue uniform, unless it was a white Yankee officer leading the colored troops. If captured, white USCT officers were to be tried for inciting a slave insurrection, punishable by death, according to Southern courts.

The Southern government did not condone execution of prisoners, but the practice was not uncommon. On December 20th, two white officers of the 12th U.S. Colored Troops and one from the 44th USCT were captured by the Forrest scout company of Capt. Addison Harvey about 14 miles southeast of Murfreesboro. They were stripped of their uniforms, and marched for the better part of two days. They were then directed into a ravine three or four miles west of Lewisburg and each man shot in the head. One miraculously survived—Lieutenant George W. Fitch of the 12th USCT. He was shot behind the ear with a pistol, a glancing blow, the bullet lodging in the bone behind the ear. He pretended to be dead and managed to crawl to a nearby house, surviving the horrendous incident.

By nightfall on Tuesday, December 20th, Wood's Federal infantry rested on the north bank of the Duck River, opposite Columbia, waiting for the pontoon train to arrive. Wilson's cavalry remained in bivouac, waiting for supplies, while Hatch's men stared across the river at Columbia. The main body of the Confederate army was well on its way down the turnpike to Pulaski (where Hood had already established his headquarters), while Forrest's cavalry rearguard and Walthall's infantry lingered at Columbia.

As the Federals fumed on the north bank of the Duck River, Hood established his headquarters in Pulaski at the home of Thomas Jones. From Pulaski, it was 49 more miles over abominable roads to the Tennessee River at Bainbridge. Most of Hood's army would reach and pass through Pulaski in the next two days.

Acting on a directive from Thomas, who was using every resource at his disposal, beginning December 20th, Acting Rear Admiral Samuel P. Lee led a flotilla of gunboats up the Tennessee River towards Florence, where the Confederates had established an artillery battery. Thomas was hoping that the gunboats, in a naval flanking maneuver, could destroy any bridge over the Tennessee so that the Federals could bag Hood's army on the north bank. Lee's squadron boasted more guns than Lt. Commander LeRoy Fitch had commanded at Nashville, and included the river monitor *Neosho*, the City Class ironclads *Carondelet* and *Pittsburg*, and the timberclad *Lexington*.

## Desperate Desolation
*Wednesday, December 21st, 1864*

By daylight on Wednesday, December 21st, snow covered the ground, with man and beast suffering from the bitterly cold temperatures. Hood's army began moving out of Pulaski, heading southwest on Lamb's Ferry Road, with corps commander A.P. Stewart now assuming a major role in leading the main body of the army. "Stewart unquestionably had a significant role in directing the retreat, as dictated by his being the army's second-ranking officer and having better mobility than the disabled Hood," according to historian Elliott.

Capt. Gale of Stewart's Corps recorded the miserable conditions: "On we marched, through ice and rain and snow, sleeping on the wet ground at night. Many thousands were bare-footed and actually leaving the prints of blood upon the ground. Enemy pressing us in the rear. When we left the pike at Pulaski, we had an awful road, which was strewn with dead horses and mules, broken wagons, and worse than all, broken pontoons. How we counted them as we passed them, 1, 2, 3 to 15." Those were the pontoons needed to cross the Tennessee River to safety. There was no turnpike now, only country roads that wound through the increasingly hilly terrain, a perfect setting for an ambush.

James D. Porter, chief of staff and assistant adjutant-general, Cheatham's Corps: "We had reached the hilly country in Giles County, beyond Pulaski. It had snowed and sleeted the day before, and the ground was as slick as glass. We reached a steep hill, and I rode on to the top with the troops. General Cheatham remained at the foot of the hill, and he knew they were going to have terrible times with that train of his approaching with ordnance stores, quartermaster's stores, etc. He sent word to me to pick out a hundred well-shod men and send them to help push the wagons up. I dismounted and gave my horse to the courier. The fellows soon found out that I was after men with shoes on, and they were highly amused. They would laugh and stick up their feet as I approached. Some would have a pretty good shoe on one foot and on the other a piece of rawhide or a part of a shoe made strong with a string made from a strip of rawhide tied around it, some of them would have all rawhide, some were entirely barefooted, and some would have on old shoe tops with the bottoms of their feet on the ground. I got about twenty or twenty-five men out of that entire army corps, and we got the teams up the hill."

Sgt. James R. Maxwell of Lumsden's Alabama Battery: "I had a sort of horse to ride, but could not ride because of my boils. My boot soles were held to the uppers by strings and wires. My socks were worn out. The roads were an icy, gritty slush, from the sleet that fell almost daily. I could only hobble. So, to keep up, I would watch for a chance and crawl into any wagon possible and stay there all day... It was steal rides in a wagon and keep up, or I might be captured."

Hundreds, if not thousands, of horses and mules were succumbing to exhaustion and exposure to the elements. Once consumed, the beasts were of no further use and could only suffer. If an animal's legs or hoofs gave way from exertion, the horse or mule was shot and left by the side of the road, along with all the other debris.

U.S. General John Beatty spoke at Christmas 1863 at a Chattanooga gathering on the subject of war horses: "There is no suffering so intense as theirs. They are driven with whip and spur, on half and quarter food, until they drop from exhaustion and are abandoned to die in the mudhole where they fall ... A man can give vent to his sufferings, he can ask for assistance, he can find some relief either in crying, praying, or cursing; but for the poor exhausted and abandoned beast there is no help, no relief, no hope."

Before advancing into Tennessee, Hood had inventoried 124 field artillery pieces, but by December 21st his army had only 59. Two pieces of McKenzie's Alabama battery were left at the Duck River because the pontoon bridge had already been removed. Many of the army's batteries were wiped out as fighting units: Douglas of Texas; Dent's Confederate; Stanford of Mississippi; Lumsden, Tarrent, Eufala, and Selden of Alabama; the Louisiana Washington Artillery; and Fenner of Louisiana. In the famed Washington

Artillery, Colonel Robert F. Beckham and Capt. John Rowan had been killed, and Major Daniel Trueheart captured. After the battle at Nashville, only seven of their field officers remained on duty.

Meanwhile, more than 30 miles to the north, the Federal pontoon train rambled over cross-country roads to get to the Columbia Pike (from the Murfreesboro Pike) and finally caught up on the evening of December 21st, five days after the pursuit had begun. A portion of the pontoon train arrived at Rutherford Creek about 1:00 pm. The train was dismantled to bridge the stream so that the remainder of the pontoon train could proceed to the Duck River. A thousand troops from Schofield's XXIII Corps were sent to help with the bridge. Thomas discovered on the evening of December 21st that due to the weather the pontoon bridge would not be completed over Rutherford Creek before nightfall. As a result, work on laying the Duck River bridge could not begin before daylight on December 22nd. By nightfall, Wood got all of his troops over Rutherford Creek, then A.J. Smith's corps crossed the stream.

With all the delays, it would seem that Hood's army was rapidly slipping away. Once across the Duck River, however, it was assumed that Wilson's refreshed and well-armed cavalry would gallop down the turnpike posthaste and easily reach Hood before he escaped. Late on December 20th, General Thomas sent an order to Wood, which read in part: "It is the desire that the entire army be over the (Duck) river before tomorrow night, in which case it is to be hoped that the greater part of Hood's army may be captured, as he cannot possibly get his teams and troops across the Tennessee River before we can overtake him."

On December 21st, Sherman's march to the sea ended as his troops entered Savannah, Georgia on the East Coast. Sherman telegraphed Lincoln, "I beg to present to you, as a Christmas gift, the city of Savannah..." Sherman had led 62,000 troops 285 miles across Georgia and cut a path of destruction more than 50 miles wide. The Confederate opposition, led by Hardee, retreated, and crossed over the Savannah River into South Carolina. Sherman's Carolina Campaign, destined to be much more destructive than his march through Georgia, would soon begin.

## Over The Duck
### *Thursday, December 22nd, 1864*

On Thursday, December 22nd — Brigadier General Ed Hatch's 32nd birthday — work began at 5:00 am on the pontoon bridge across the Duck River at Johnson's Landing, one or two miles above Columbia, under the direction of the energetic Colonel Abel Streight, a Hoosier. Streight boasted a colorful military career. In 1863 he led an ill-fated raid into the Deep South with his men mounted on mules. He was chased by Forrest, who outfoxed him and persuaded him to surrender his command. In other later action, Streight was captured and sent to the notorious Libby Prison in Richmond, where he and others escaped by tunneling out.

It was quickly learned that there were only three experienced pontoniers with the entire Federal bridge-building team. At 7:00 am, Forrest's men began firing on the construction party from across the river. The brave men of the 51st Indiana somehow managed to paddle across the river in canvas pontoon boats and establish a bridgehead. It required all day to fashion a rickety, poorly secured structure. Two or three times, the partially completed bridge broke apart. The rain and snow had stopped, but it was bitterly cold, the temperature dropping to 15 degrees. The river current brought along chunks of floating ice which hampered the workers. The structure was completed at 6:30 pm, and Wood's men crossed during the night. Wood reported in routine understatement that "owing to the inexperience of the troops in such service and the extreme coldness of the weather, more time was consumed in doing it than could have been desired."

Thousands of U.S. infantry began crossing the Duck River. The winds were out of the north, "whistling like a zephyr across the frozen landscape," according to one witness. At 9:00 pm, the

1st Division of Wood's IV Corps crossed the Duck River in the darkness and bivouacked about midnight south of Columbia. Six miles down the Pulaski Pike was all the Federals could move that day. Several hundred Yankee infantrymen had crossed when Forrest directed Walthall to start the rearguard infantry on the pike south toward Pulaski.

Pulaski, a small town (pop. 1,500) and the seat of Giles County, was named for Count Casimir Pulaski, a Polish nobleman, hero of the American Revolutionary War, and acknowledged as the father of American cavalry. He wrote a manual of cavalry tactics. He died at the age of 34 from wounds suffered at the 1779 Battle of Savannah, Georgia. He is one of only eight persons to be awarded honorary United States citizenship. After the war, the small town became known as the birthplace of the Ku Klux Klan, with Forrest serving as its first grand wizard, and famous for its statue honoring Sam Davis, the "boy hero of the Confederacy."

Forrest withdrew toward Pulaski, leaving Chalmers on the Bigbyville Pike to the west, and Buford and Jackson protecting the infantry on the main pike. Forrest dropped back three miles down the pike to Warfield's Station, where skirmishing broke out. Outflanked, Forrest fell back 12 miles farther and at nightfall went into camp at Lynnville (the current site of the village of Waco; after the war Lynnville relocated a short distance to the east, on the railroad).

A few miles farther south, the Washington Artillery was ordered to hand over all its mounts for use with the Confederate pontoon train. Their caissons and limbers were pulled onto the bridge over Richland Creek and all the ammunition was dumped into the water. The empty caissons and limbers were pulled back to camp, piled up, and burned. It should be noted that Richland Creek, much like Rutherford Creek, consisted of many branches. There was a turnpike bridge just south of Buford's Station and a larger, covered bridge just outside Pulaski.

Dr. W.J. McMurray wrote in his history of the 20th Tennessee: "To show the spirit, wit, and fun there was in the Confederate soldier, while half-clad and half-starved and barefooted, and fighting three to one, I will relate this: on the retreat near Pulaski, the roads were muddy and crowded, and every soldier was pulling along as best he could. General Hood and staff were passing, and as they were about to crowd an old soldier out of the road, he struck up this song, where General Hood could hear it (sung to the tune of *Yellow Rose of Texas*):

*My feet are torn and bloody,*
*My heart is full of woe,*
*I'm going back to Georgia to find my Uncle Joe.*
*You may talk about your Beauregard,*
*You may sing of Bobby Lee,*
*But the gallant Hood of Texas*
*He played hell in Tennessee.*

Hood's headquarters were made at Pulaski the night of December 22nd. Stevenson's Corps (formerly S.D. Lee's) was directed to move forward on the Lamb's Ferry Road, in rear of the pontoon train, and camped about eight miles from Pulaski. A.P. Stewart's Corps camped in the rear of Stevenson's about six miles from Pulaski, and General Cheatham's on Richland Creek, in the immediate vicinity of town. The wagon train was ordered to move at daylight toward Bainbridge, by the Powell Road.

### Blunting The Pursuit
*Friday, December 23rd, 1864*

On December 23rd, the temperature continued to drop, and the Federals struggled to get across the river. At 5 am, General Wood assessed the situation at the Duck River — "the bridge is in such a bad condition and the descent and ascent of the banks so slippery that it is most difficult to get on and off of the bridge. Since midnight, when the last of General Elliott's division crossed, we have been able to cross but three batteries and a few wagons. The rest of our artillery and the greater part of our train is to cross, but the bridge must now be given up to the Cavalry Corps, which is just ready to cross." Nine hours later, Wood noted that "the cavalry is very slow crossing the bridge."

It took Wilson's cavalry, resupplied and rested, the entire day to cross the Duck River, Wood's IV Corps having already crossed. Tired of waiting for the troopers, Wood began the march south at 2:30 pm, and 90 minutes later the infantry confronted Forrest's pickets at the Warfield plantation site and drove them farther south with a volley of artillery fire. A short time later, the Federal infantry came upon Forrest's men deployed in a gorge between two high hills five miles south of Columbia. The Federal infantry pressed forward on the turnpike while the cavalry, now caught up, pressed on the flanks despite the fact that the secondary roads were described as nearly impassable.

Forrest's account of the day's events: "The enemy made his first demonstration on my rear pickets near Warfield's, three miles south of Columbia. He opened upon us with artillery, which forced us to retire farther down the road in a gap made by two high hills on each side of the road, where he was held in check for some time. On the night of the 23d I halted my command at and near Lynnville, in order to hold the enemy in check and to prevent any pressure upon my wagon train and the stock then being driven out."

Wood's account regarding the IV Corps and the affair at the gap between the hills: "After advancing some five miles south of Columbia, the afternoon of the 23d, the head of the corps came on a party of the enemy posted advantageously in a gap through which the highway passed, with inclosing heights on either side. I ordered Brigadier-General Kimball, commanding the leading division, to deploy two regiments as skirmishers, bring up a section of artillery, and with this force to advance and dislodge the enemy from the pass. The service was handsomely and quickly performed."

Wood reported that one Confederate cavalry captain and one private were killed, and four infantrymen captured. Being close to nightfall, Wood halted his corps to wait for all of the Federal cavalry to cross the Duck.

Federal division commander Brig. Gen. Nathan Kimball reported: "The advance guard of cavalry immediately in my front came up with the rear guard of the enemy about five miles south of Columbia, strongly posted in a pass between high hills and through which the road ran. I immediately deployed a strong line of skirmishers and sent them forward. A section of Thomason's (First Kentucky) battery was put in position about 800 yards from their lines and opened upon them. After a sharp skirmish they were driven from the pass, leaving behind a captain mortally wounded and one man killed. My command bivouacked for the night in the pass."

Forrest was employing a tactic that Daniel Morgan had successfully employed at the Battle of Cowpens during the Revolutionary War. The Confederate cavalry commander would station a forward line on the pike and then position the main body of the rearguard, supported by artillery, at a second fallback site. When the Yankees made contact, the first line would fire a volley and then fall back, drawing the Federal cavalrymen into what was, in effect, an ambush. This tactic worked well in delaying the Federal pursuit. Col. Charles Olmstead of the 1st Georgia infantry said the tactic "compelled the Yankees to halt, reconnoitre, and deploy and then feel our lines before advancing to the attack." This took time, and "time was what we were fighting for," said Olmstead.

Federal cavalry leader Wilson agreed: "The country on the right and left of the pike, very broken and densely timbered, was almost impassable; the pike itself, passing through the gorges of the hills, was advantageous for the enemy; with a few men he could compel the pursing force to develop a front almost anywhere."

Private Nelson Rainey of Confederate General W.H. Jackson's cavalry division rode with the rearguard and described the conditions: "The weather was very cold…rain, half sleet, then snow half sleet on the rocky frozen roads. We all suffered. The infantry most of all. Not half of these poor boys had blankets, very few overcoats. More than half without shoes, their feet tied up in gunny sacks or old cloth. We have all read in history that Washington's barefoot soldiers left bloody tracks on the ground. I saw such instances, plenty of them, on this retreat. The boys were hungry too, all hungry. At one place, our company commissary officer, Bill Eanes, found a pen of fairly fat hogs. We had a day's ration of pork which we ate raw—everything too wet to make a fire. At a cabin I parched corn in an old shovel. (A comrade) and I lived on that for two days."

During the night of December 23rd, one week after the debacle at Compton's Hill, much of Hood's army reached the proximity of Lexington, Alabama, about two-thirds of the way from Pulaski to the river. Army headquarters was located on Powell's Ferry Road, six miles from Lexington. The army, after the day's march, camped as follows: Stevenson's corps at the intersection of Lamb's Ferry Road and Powell's Ferry Road, four miles from Lexington; Stewart in the rear, on the Lamb's Ferry Road; and Cheatham on the Lawrenceburg Road.

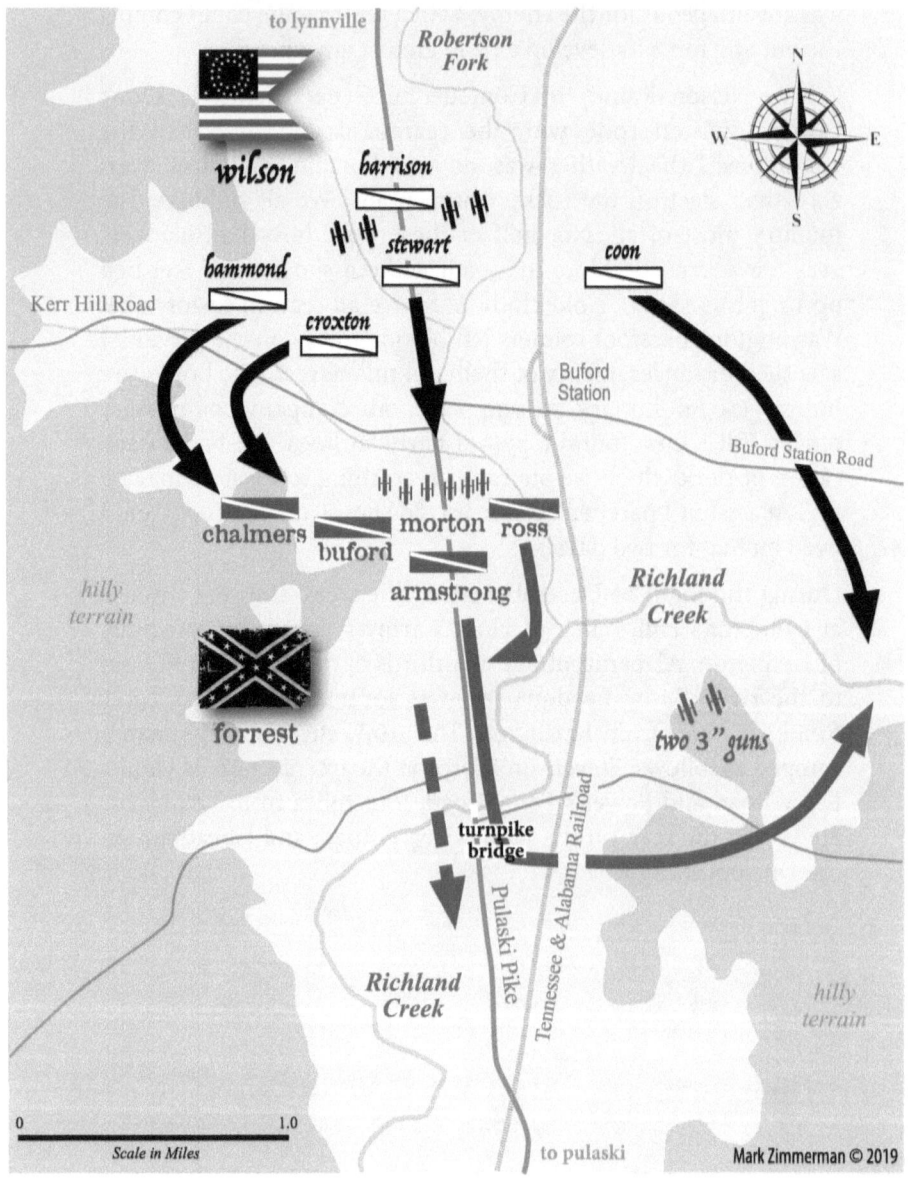

## Richland Creek
*Saturday, December 24th, 1864*

By early dawn, Saturday, December 24th, the last of the Federal Army reached Columbia on the south bank of the Duck River. By 7:00 am, Wilson's cavalrymen began trotting past Wood's infantry on the turnpike, assuming the lead. At noon, at long last, the Federal pursuit began again in earnest. After five days of delay, the U.S. army was in full pursuit. It would be a day of constant skirmishing, accentuated by periods of heavy fighting.

John R. Scales, retired U.S. Army brigadier general and author, wrote about the tactics of the Federal cavalry commander: "Wilson had five strong brigades, three independent (their division commanders and other brigades were not present) and two brigades organized into a division under Hatch. During the three days when Wilson led the pursuit, he used a different independent brigade as his lead or point element each day, thus rotating the unit exposed to the most danger. He kept Hatch's division back in a column as an integrated unit. So he had a massive force that could maneuver as a whole once initial contact was made."

Wilson was the overall Federal cavalry commander. Coon and Stewart were in Hatch's command, while Croxton, Harrison, and Hammond operated as the three independent brigades.

About two and a half miles from Lynnville, Walthall's rearguard infantry occupied an advantageous position between two large hills. This position was held until sunrise on December 24th, when the retreat was resumed.

At Lynnville, Forrest decided to put some distance between the Federal cavalry and his wagon train. He sent the rearguard north three miles up the Pulaski Pike near the Bigbyville Road to confront the enemy, with Walthall's infantry on the pike and cavalry on both flanks. The result was a "severe two-hour engagement." By this time, Chalmers had left the Bigbyville Pike to the west and joined the main body of cavalry. He was attacked by Hatch's division and Croxton's brigade, then fell back to join with Buford. Forrest determined to make a stand on the pike at Richland Creek, near the Buford railway station, about eight miles north of Pulaski. Walthall withdrew his infantry to Richland Creek to hold the crossing for the cavalry if they should be hard pressed and need to cross before nightfall. He stayed there until dark on December 24th, then withdrew seven miles to the outer line of earthworks near Pulaski.

After the brisk firefight at Lynnville, the next major obstacle to the Federal advance was Richland Creek, the turnpike bridge about seven miles north of Pulaski. The countryside near Richland Creek flattened out, with an absence of steep hills. There the two cavalry forces (no infantry was involved) clashed in a desperate fight.

Under counterattack at Lynnville, Forrest retreated several miles to a position just north of the Richland Creek bridge. He placed Armstrong's brigade of Jackson's division on the pike in support of Morton's six smoothbore artillery pieces. The troopers under Sul Ross, Jackson's other brigade, were placed to the right. Chalmers and Buford were situated to the left of the pike. Morton's best artillery, a pair of three-inch rifles, were sited on a bluff southeast of the bridge.

Croxton led Wilson's cavalry charge down the pike that day. He was ordered to swing around the Confederate left flank, along with Hammond's command. Stewart's brigade of Hatch's division was sent straight down the pike, with Harrison's troopers back on the pike in reserve. Coon's brigade was sent on a wide swing around the Confederate right flank. About 6,000 Federal cavalrymen and 12 guns faced some 3,000 Confederate cavalrymen with eight artillery pieces.

A brisk artillery exchange commenced, with Forrest claiming to have dismounted two of the Federal guns. Meanwhile, the 6th Texas Cavalry, backed by the 3rd Texas, led Sul Ross' men in checking the Federal advance down the pike. Coon managed to cover much ground but was thwarted by the lack of a ford across the swollen creek. In response to the flanking maneuver, Forrest sent Ross and Armstrong back down the pike, across the bridge, and east to meet Coon's threat. At Richland Creek, Morton's guns delayed the Yankee troopers long enough for the Confederate gunners to repair the damage to the bridge caused by the earlier passing of Hood's troops. Morton's cannons, called Bull Pups by his men, were then carried across by hand, the horses led over, and the planks pulled up and thrown into the creek.

Croxton drove elements of Forrest across Richland Creek, capturing a few prisoners. Corporal Harrison Collins of Co. A, 1st Tennessee (U.S.), captured a guidon from Chalmers' command, for which he later received the Congressional Medal of Honor. During this maneuver, Croxton accused the 8th Iowa of repeatedly failing to respond to orders to cover the right flank. Such failure prevented them from capturing the rebel artillery and many prisoners, Croxton claimed.

Capt. Obediah Hayden of the 9th Indiana cavalry, which was not engaged, spoke of the sights along the pike on December 24th: "All this day, as we followed in the wake of the fight (on the right of Croxton's brigade), our eyes were constantly greeted with unmistakable evidences of the struggle in front. No one will forget the little knot of dead and dying artillerymen and horses by the road-side, maimed and mangled by a bursting shell, a gory, ghastly sight."

Abe Buford was severely wounded (near Buford's Station, ironically, which was not named for him), and Forrest ordered a retreat when Croxton's and Hammond's brigades flanked Chalmers' line. Forrest continued all the way to Pulaski, and though Wilson pursued, he did not catch up that night. In the fighting along Richland Creek, after Buford was wounded, Chalmers assumed command of his unit and merged it with his own. Both commands were about the size of a brigade.

From Richland Creek to Pulaski, fighting was mostly hand-to-hand. Buford took a flesh wound in the leg from a Spencer carbine round. Tyree Bell was put in command and didn't know the location of the Kentucky troops, commanded by Lt. Col. Absalom R. Shacklett of the 8th Kentucky Infantry (mounted). Bell finally did locate them and ordered them across the creek. They were the last across before the bridge was demolished, leaving Bell and part of his escort on the north side. An artillery shell burst knocked Bell from his horse and blinded him in his right eye, permanently

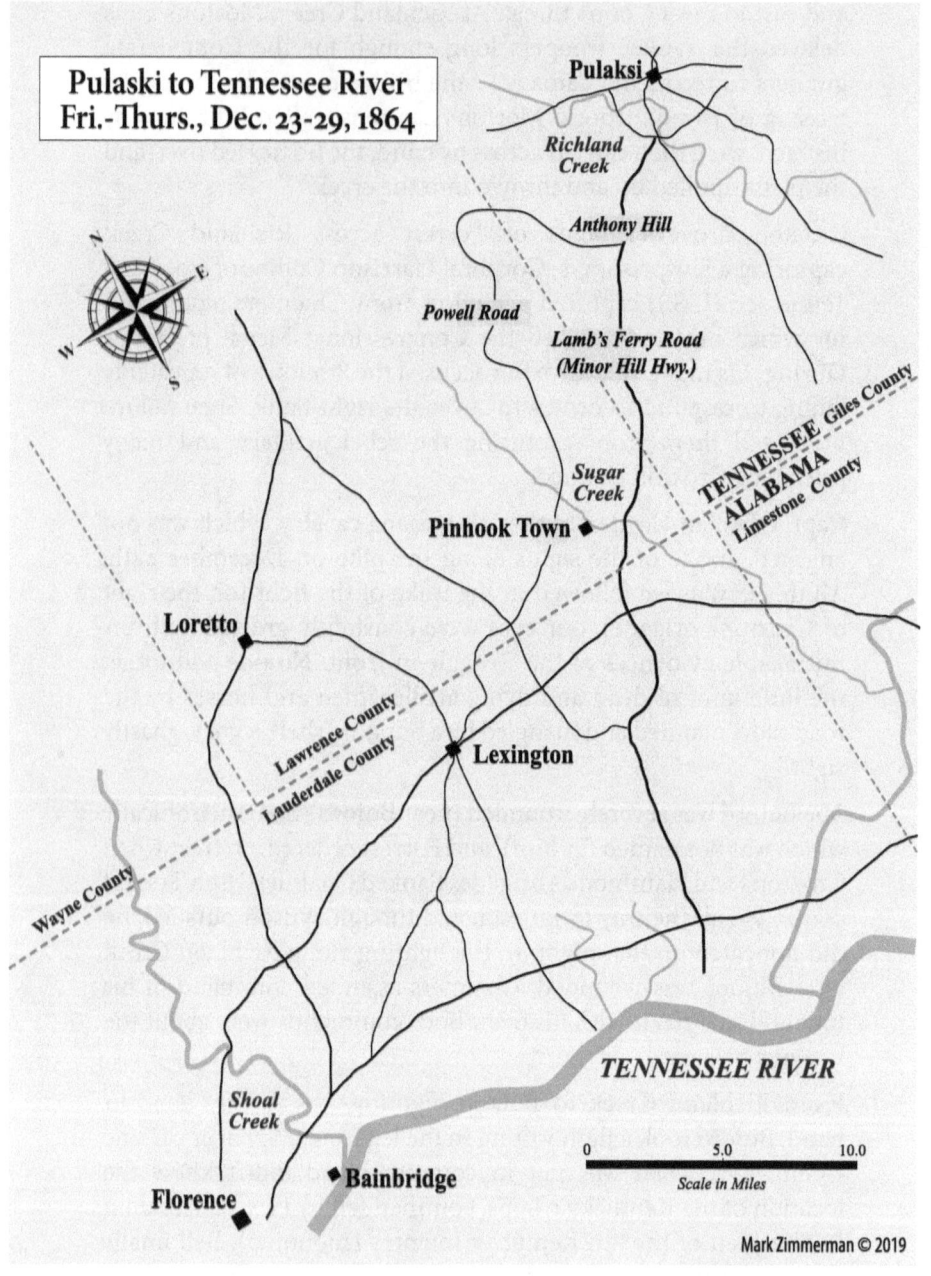

as it turned out. Chalmers was also there on the north bank with some of his men. They all fled for their lives, trying to hide among the hills. One Federal cavalryman nearly reached Chalmers with his saber before he was shot by Lt. Col. William F. Taylor of the 7th Tennessee Cavalry. They rode to the top of a hill, turned, and volleyed into their pursuers, staggering them. They went down the hill, up another, and did the same thing. They raced across a cornfield to the creek bottom, and for some reason the Federals did not contest their movements. They swam their horses to the south bank of Richland Creek and greeted Forrest and Walthall. Forrest lost one killed and six wounded. After nightfall, Forrest reached Pulaski "without further molestation."

Bell refused to dismount and retire, despite great pain in his eye. Forrest ordered him to turn over his command, and the two of them rode to Pulaski to enjoy a supper at Tom Martin's house. Once there, Dr. J.P. McGee ordered Bell to go to the hospital, but Bell again refused. He asked for a face wash, plaster to stick up the cuts, a clean shirt, and then a good supper. The Yankee cavalry rode into town as they were finishing their meal. They fled through the covered bridge over Richland Creek and pickets set fire to it. The Federals put the fire out and patched it up. Forrest and Bell took to the woods, eluded them, and made their way back to camp.

One cavalryman under Forrest described the brutality of the fighting during the retreat: "The enemy's cavalry swooped down upon us with drawn sabers, cutting and slashing us from right to left. Three soldiers assaulted General Buford at one time. One he shot; another he struck over the head with the butt of his pistol; and the third he grabbed by the hair and pulled from his saddle and thus escaped. They swarmed around me like a flock of blackbirds. How I got out of it with a whole skin, I do not know. My face was powder-burned and my hair was scorched from a pistol shot thrust in my face at the moment of discharge, and I found myself with two severe bruises on the shoulder from saber strokes."

In the main body of the Confederate troops, artilleryman Private Stephenson continued his retreat on foot. "Now my rags were about worn out, and my feet, torn, bleeding, swollen and frost bitten." He entered a little house with a light burning. A young mother with children sat in the house. He asked for help and she gave him a pair of pink children's socks. He wrapped them around his feet like bandages and continued onward.

South of Pulaski, the countryside seemed to become a barren wilderness, a no-man's land. Wandering off the main route to forage was dangerous. Pvt. Stephenson noted: "We soon left behind us Middle Tennessee. Before us lay the wild, barren, mountainous tract that makes the southern border of Tennessee a land of bushwhackers and outlaws. Treachery, ambush, assassination lurked on every side, even within a hundred yards of the road."

Some of Hood's men moved into Alabama, passed through the small town of Lexington, and reached Shoal Creek, the last obstacle before reaching the Tennessee River, on Saturday, December 24th, when there was four inches of snow on the ground. The pontoon bridge was being built at this point, near the small village of Bainbridge, where the river was wide and the current strong. The threat of Federal gunboats steaming upriver to shell the pontoon bridge occupied everyone's thoughts.

Guarding the procession were riflemen, stationed near the river at a deep, narrow ravine barely wide enough to accommodate the steady stream of wagons rumbling southward. On the northern bank of the Tennessee, the pontoon bridge rested on a narrow strip of land formed by the mouth of Shoal Creek, which terminated almost parallel to the mighty river. Would there be enough pontoon-boats to reach the opposite shore? — the engineers still didn't know for sure.

The ragged men were hesitant to cross Shoal Creek in the icy conditions. But after General William F. Brantley's horse slipped in the icy creek and dumped him, the reluctant men jumped into the creek to cross. Pvt. Stephenson: "The water was very, very cold, but there was a row of fence fired for us to warm by on the south bank." They spent the night of December 24-25th "drenched, like drowned rats from the icy waters of that creek, men huddled together on that little strip of land, waiting, assured that the pontoon bridge would be ready in the morning." (The Washington Artillery, Pvt. Stephenson's unit, finally crossed on December 26th, taking 10 days to march 120 miles from Nashville.)

George E. Estes of the 14th Mississippi, who had been wounded and captured during the war, recalled: "We met no obstacles until we came to Shoal Creek, where we were halted and ordered into line of battle to hold the Yankees in check till our army could get across the creek on an improvised bridge made of fence rails. A (square) rail pen was built in the middle of the stream by placing rail upon rail and rail upon rail until the pen was above water, and then a man would get on each corner and hold down the pen till stringers could be placed from the bank reaching into and upon the rails. After this, they would cover the stringers crosswise till a thing somewhat resembling a bridge had been constructed. General Featherston was in command and he was trying to get the men in some sort of order to resist the Yankee cavalrymen. He would put stragglers in the line and go off to get more. When he came back the first band had wandered off, so he gave up on any formidable resistance. The Yankees did not attack. We passed on through the mud and slush until we finally came in sight of the Tennessee River, with its wide expanse of murky waters bearing down the stream drifts of logs, trees, brush, and everything imaginable."

On December 24th, U.S. Admiral Samuel Lee arrived at Chickasaw, Alabama, on the Tennessee River with four of his five Federal gunboats. After steaming about two miles upriver from Florence, the gunboats encountered a Confederate battery of two Parrott guns on the north shore protected by earthworks. Volleys were exchanged with little results, and the Federal gunboats fell back downstream.

The night of December 24th, Hood made his army headquarters at the Joiner house, eleven miles from Bainbridge, on the main Bainbridge road. Stevenson was encamped on Shoal Creek, with Stewart right behind. General Cheatham's Corps had not yet come onto the main road from the Powell Road.

That night, General Thomas received a message from Secretary of War Stanton in Washington. President Lincoln had nominated Thomas, "that Virginian," for rank of major general in reward for his victory at Nashville. The telegram stated: "...no commander has more justly earned promotion by devoted, disinterested and valuable services to his country." An exasperated Thomas told Chief Surgeon George E. Cooper, "I suppose it is better late than never, but it is too late to be appreciated. I earned this at Chattanooga."

## Christmas Day
### *Sunday, December 25th, 1864*

Sunday was Christmas Day 1864, but there was no holiday truce. Except for a few impromptu services, few of the men in the vicinity of Pulaski, Tennessee felt like celebrating. Certainly, many felt the need to pray. Those clad in gray prayed for salvation, and a bridge across the wide Tennessee River upon which to cross to safety. The Federals most likely prayed for a little more time and a lot more luck. And better weather.

Over in northeastern Mississippi, on Christmas Day, Federal cavalrymen under famed raider Brig. Gen. Benjamin Grierson struck Verona, a key Confederate supply depot south of Tupelo. Before leaving, the Federals burned or otherwise destroyed two locomotives, 32 rail cars, eight warehouses, and 200 wagons, all supplies intended for Hood's army. The Federal cavalry also severed Hood's lone railroad supply line between Corinth and Tupelo.

Back at Franklin, the wounded were being transported to better facilities in Nashville. Dr. Deering Roberts, the surgeon of the 20th Tennessee, said that "... my associates and myself, with the wounded of Bate's division, were all moved to Nashville, and placed in the large building on South College Street, where I, as the ranking surgeon, assumed charge of the 1,200 wounded there assembled from the battlefields of Franklin and Nashville." Dr. Roberts was referring to the newly built Howard School, a three-story brick building with an Italianate clock tower.

Christmas Sunday, nine days after Compton's Hill, found the main body of Hood's army reaching the Tennessee River and beginning to fashion a pontoon bridge across the broad expanse of rising water. At the same time, the Confederate rearguard was clearing out of Pulaski as the Federal cavalry arrived with their infantry close behind. In many ways, the pursuit and the retreat would become even more desperate the closer the competing forces moved toward the river. In addition to looking over their shoulders for Federal cavalry, the Confederates worried if there were enough pontoon-boats to span the river and whether the Federal gunboats would blow the bridge to smithereens. One direct hit might do the trick. Due to the rainy weather, the river level was up and many officers believed that the odds of U.S. gunboats reaching the pontoon bridge "not improbable."

Approaching the Tennessee River, A.P. Stewart's aide, Colonel Gale, wrote to his wife: "Every man was haunted by the apprehension that we did not have boats enough to make a bridge."

As Stewart's men improvised earthworks along Shoal Creek, Hood's chief engineer, Lieutenant Colonel Stephen W. Presstman, got busy building the pontoon bridge at the village of Bainbridge, just below Muscle Shoals and four miles upstream from Florence. The pontoon train was commanded by engineer officer Capt. Robert L. Cobb, who realized that he didn't have enough pontoon-boats to span the river. What he found at Bainbridge was nothing short of a miracle, however. On their way north a month ago, the Federals, for some reason, had abandoned 15 pontoon-boats at Decatur, Alabama. On December 19th, Confederate Brig. Gen. Phillip D. Roddey's cavalry came to the rescue and floated the pontoon-boats down the river to Bainbridge. But would those additional boats be enough? The engineers would soon find out.

Cheatham's Corps formed a line protecting the bridge from the west while Stevenson's corps formed to the east. The bridge was built as fast as the pontoon wagons arrived at the site.

Robert A. Jarman of the 27th Mississippi, Brantley's Brigade, described the procedure: "The first thing to do was to lash two or three pontoon boats together and use them as a ferry boat to cross over some artillery and horses to go towards Florence and protect our bridge from Federal gunboats until the army could cross."

The bridge was built in sequence, ideally with six crews performing six specific tasks, as needed. First, the approach to the bridge had to be secured with cables and abutments built up with dirt and stone. A pontoon-boat (about 26 feet long) was secured in the waterway with anchors upstream and downstream and lashed with the balks (5x5-inch beams) that would link the pontoon-boats and carry the roadway planks. Each pontoon-boat ideally had an upstream anchor, and every other boat had a downstream anchor. The pontoon-boats were usually spaced 20 feet apart, although the spacing could vary, as required. Once the balk crew had lashed down the balks and pontoon-boats, the "chess" crew would lay down the planking, to form a roadway 13 feet wide. Then, finally, another crew would place the side rails, which acted as curbs, preventing wagon wheels from rolling off the planks. Most all of these bridge pieces were made to fit together in a certain way, facilitating the construction. Boat after boat, the bridge came together and spanned the wide river, bowing noticeably in the middle with the current.

"We could see a rickety pontoon bridge hastily and insecurely built," recalled George Estes of the 14th Mississippi. "It was serpentine in shape, about twelve feet wide and half a mile long, covered from end to end with all kinds of beasts, wagons… There was a Yankee gunboat half a mile or more down the river throwing bombshells, trying to break the bridge; but the shoals prevented it from getting near enough to do any damage. The north end of the bridge had a promiscuous mass of humanity and animals all trying to get on and over the bridge at the same time. Finally my time came, and I went forward in great fear that the cable would break and let us all go down into eternity together. But the Lord… permitted us to reach the southern shore without the loss of one."

Even with the pontoon bridge completed, the Confederates' worries were not over. At this point, there was some speculation as to whether the Federal gunboats would be able to paddle upstream far enough to damage or destroy the Confederate pontoon bridge. Acting Rear Admiral Samuel P. Lee's flotilla of gunboats were approaching Florence and the dangerous shoals farther upriver, near the Confederate bridge. Lee's squadron boasted more guns than Lt. Commander LeRoy Fitch had at Nashville, and included the ironclads *Neosho, Carondelet, Pittsburg,* and the timberclad *Lexington.* Even with the rising waters, the ironclads *Carondelet* and *Pittsburg* had to be left behind at Eastport, Mississippi, while the *Neosho* river monitor and the tinclads *Reindeer* and *Fairy* engaged with two four-gun Confederate batteries at Florence, consisting of elements from Capt. James A. Hoskins' Mississippi battery, Cowan's (Lt. George H. Tompkins') Vicksburg battery, and

Phillips' Tennessee battery. The Confederates stopped firing after 30 minutes and the gunboats drifted back downstream. The next day another futile engagement resulted in the *Neosho* being hit 27 times with little damage, and the Federal flotilla suffering three killed and five wounded.

Before dawn, back in Pulaski, Forrest was busy destroying supplies left behind by the main body of Hood's army. He ordered Walthall's infantry seven miles south to Anthony's Hill (also known as King's Hill), an advantageous site to stage an ambush. The cavalry followed the infantry, save for Jackson's division, which was ordered to destroy the military supplies and burn the covered bridge over Richland Creek just south of Pulaski.

Colonel Thomas J. Harrison's brigade formed the vanguard of Wilson's cavalry pursuit that day and got started early at 5:00 am, moving out of camp ten miles north of Pulaski. A Hoosier, Harrison, 40, had attended Wabash College in Crawfordsville, studied law, and was admitted to the bar in 1851. In 1859, he served in the Indiana House of Representatives. He was commissioned captain in the 6th Indiana infantry, and his unit was sent to western Virginia. He became part of a regiment of sharpshooters, the 39th Indiana, and was promoted to colonel in August 1861. His regiment fought at Shiloh, Stones River, and in the Tullahoma campaign. In September 1863, his regiment fought at Chickamauga and was reorganized as the 8th Indiana Cavalry. The 8th Indiana participated in the raid at Atlanta, the Kilpatrick raid in Georgia, and the Battle of Lovejoy Station.

Harrison's brigade consisted of the 16th Illinois, the 5th Iowa, and the 7th Ohio. Active skirmishing commenced two miles down the pike as elements of the Confederate rearguard (most likely Sul Ross' troopers) were encountered. The 5th Iowa drove their foes from every position and charged into Pulaski from the east, saving the covered bridge, which had just been fired, from being destroyed. Some of Jackson's troopers escaped by riding through the flaming gauntlet. Two pieces of artillery were placed and drove Red Jackson's men from the creek. The 5th Iowa suffered three killed and three wounded in this action. The Federals were then harassed by Ross' troopers as they moved down the Lamb's Ferry Road (Minor Hill Road). While in Pulaski, the Federal men captured a Confederate general—Claudius Sears, who had been wounded one week prior at Nashville and had his leg amputated. He managed to travel with Hood as far as Pulaski, but he could not travel any farther and fell into the hands of the Federals.

South of town, the Federals found 20 burning ammunition wagons belonging to Cheatham's Corps, the mules having been used to haul the pontoon wagons. Wilson reported: "They are trying to reach Florence. I will crowd them ahead as fast as possible. They are literally running away, making no defense whatever." Wilson would soon be proven wrong.

Forrest set up an ambush at Anthony's Hill, where the road swerved around the prominence. He used his infantry, cavalry, and artillery forces. Morton's guns (three to six depending on which report) were positioned on the brow of the hill, where they commanded the line of fire straight up the road. Infantry under Featherstone and Palmer flanked the guns on the hill, protected by hastily built barricades and supported by dismounted cavalry. Reynolds and Feild were held back but aligned, in reserve. Chalmers, now also commanding Buford's brigade, was on a side road, guarding the right flank. Armstrong was at the front, protecting the left flank of the main road, with Ross protecting the right flank. "So broken is the ground at that point, and so densely wooded, that there was no difficulty in effectually concealing the troops," Walthall explained.

Harrison's Federal troopers arrived about 1:00 pm to find a small body of Ross' horsemen fleeing in mock fright, trying to draw the Federals into the trap. The rebel infantry was well shielded on the heights by the dense woods. Harrison was hesitant but approached nonetheless because he thought the number of the enemy was small. He ordered his men to dismount and form into line. As they reached the critical point, Morton's Bull Pups opened with double canister and the infantry charged down the slopes from both directions. The ambush worked perfectly. Harrison's troopers broke and ran, losing about 150 men killed or wounded, according to Forrest's after-action report. Approximately 40 Confederates were killed or wounded.

Frank Smith's Battery of the 4th U.S. Artillery tried to place a gun on the road but the Federal gunmen were overrun, and the Confederates captured a 12-pounder Napoleon artillery piece, their one and only "prize" during the retreat. Harrison's retreating men jumbled with Hammond's brigade behind them, and the two

144     Mud, Blood & Cold Steel: The Retreat From Nashville-December 1864

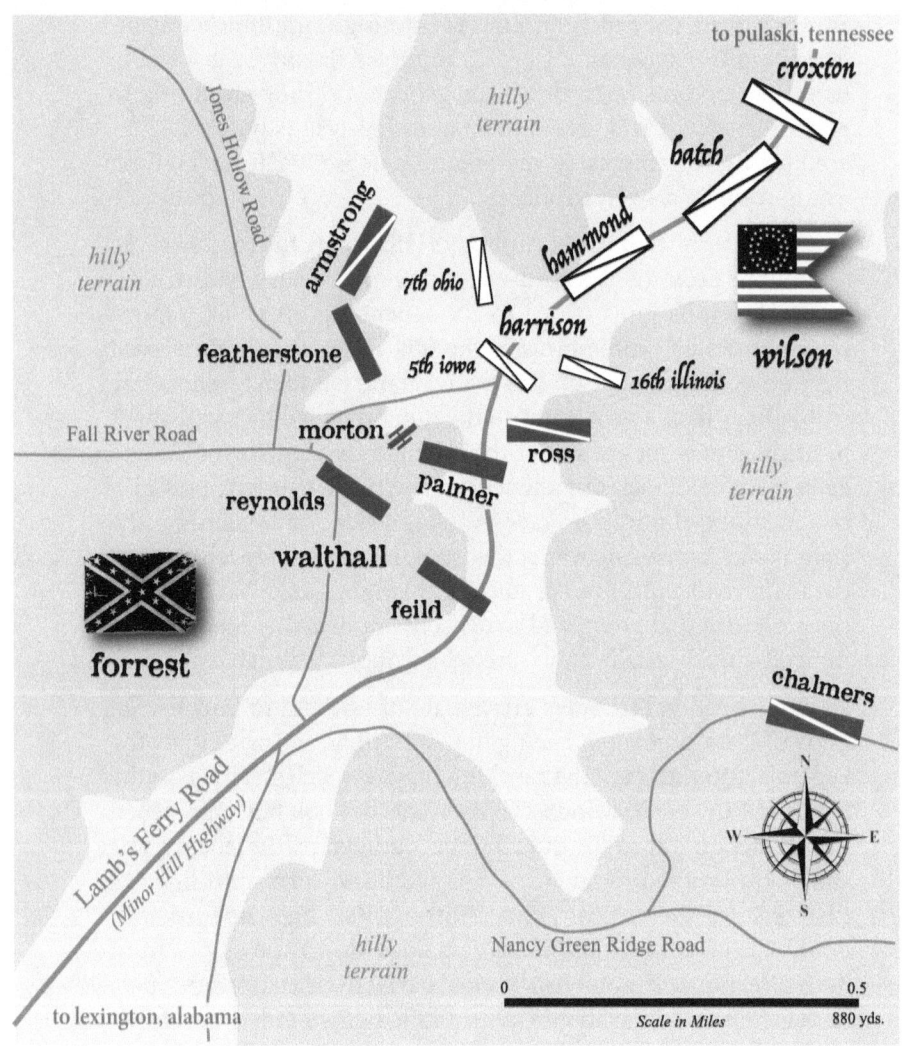

brigades fell back about half a mile to where Hatch's men were feeding their horses and waiting.

Forrest recalled his men to their original positions and stayed until almost sunset while the Federal cavalry probed both flanks. About 4:00 pm, Forrest ordered a withdrawal, and the command retreated about 12 miles to Sugar Creek, near the Alabama state line, reaching it around midnight. Hatch claimed that he counterattacked and took back the hill, but by that time Forrest's men were gone. The Federal cavalrymen encamped on the hill that night, claiming to have captured 50 of Forrest's men.

Harrison explained how his command was turned back. "(The enemy's) position was admirably selected, being hidden from view by heavy timber until within a few feet of it. Supposing that the enemy would retire from this position, as he had from others on a flank movement from us, I deployed the Seventh Ohio Cavalry on the right and the Sixteenth Illinois Cavalry on the left of the Fifth Iowa Cavalry, all dismounted. These regiments moved upon the enemy most gallantly, when suddenly he opened from a masked battery of three guns and charged over his works, in two lines of infantry with a column of cavalry, down the main road. Before this overpowering force my men were obliged to fall back about half a mile, when we checked the enemy, and, receiving support, drove him back."

Hammond, who was behind Harrison's lead brigade during the Anthony's Hill attack, reported, "The Seventh Ohio Cavalry breaking to the rear, cut my column in two just as the Fourth Tennessee, my regiment in advance, had successfully gone into action, driving the enemy into his works; and the enemy was for a time between my lead horses and the remainder of the command. I was obliged to withdraw the Fourth Tennessee to save the horses. When remounted, being joined by part of the Second Tennessee, we attacked the enemy in flank and drove him into his works again, holding the position until ordered away."

Hatch stated that at the "critical moment" he ordered the 9th Illinois regiment to dismount and charge the Confederate infantry, once Harrison and Hammond's men had fled past. The counterattack drove the rebel infantry back into their barricaded works, he stated. The whole division was then ordered to charge and the works were taken (with little to no resistance).

Capt. Obediah Hayden of the 9th Indiana cavalry described the action: "Our brigade was ordered to support the first brigade, sixth division, in an attack on this position. The enemy made strong resistance, and for some time the battle raged without advantage on either side. Hammond's brigade was ordered up. The 9th Indiana, with Companies I and D as skirmishers, advanced on the enemy's left. His skirmishers were soon met and driven back on the main line, lying along the top of the hill. A heavy fire was opened on these two companies, and they were compelled to fall back on the supporting column — which, going into line, advanced upon the enemy. Meantime the brigade in advance had been repulsed, and fell back in confusion — the 7th Ohio cavalry breaking through the advancing column between the 4th Tennessee and the remainder of the brigade. The 4th charged gallantly and drove the pursuing enemy back into his works, but the support being delayed by the demoralized 7th Ohio, failed to come up, and the 4th was compelled to withdraw. Now it was that our regiment advanced with the 2d Tennessee on our left. Coming within sight of the defenses, the whole line, with a yell and mighty rush, swept up the hill over the works and across the opening after the flying foe, who disappeared into the woods beyond. Company I had a place in this charge. Company D, being on the right of the skirmish line, had, in falling back, after helping develop the enemy's position, missed the supporting column — having to go around a precipitous hill to rejoin the regiment, only reached the scene in time to observe, but take no part in, the charge, which closed the day's work."

According to Forrest biographer Jack Hurst, as Forrest fell back he ordered Armstrong to hold his men in line despite running low on ammunition. Armstrong protested three times, riding to where Forrest and Walthall sat on their horses and said, "General Walthall, won't you please make that damned man there on the horse see that my men are forced to retreat?" Forrest replied that he was trying to gain time for Hood to cross Sugar Creek 12 miles to the south. Forrest looked at his watch and then said, "It is about time for us all to get out of here."

The Confederate troopers fell back "in darkness, rain and sleet, and bitter wind" to the "clear, pebble bottomed" Sugar Creek crossing. There they found a large portion of the ordnance train, delayed because the mules were used to move the pontoon train to the

Tennessee River. The 10th Texas cavalry, part of Ector's brigade, claimed to have killed 400 Federal horses at Anthony's Hill and captured several hundred more.

At Anthony's Hill, Wilson had sent for Wood's infantry, five miles back, to hasten forward. "We have met a slight check. There are eight brigades of infantry in our front, with rail entrenchments. Please hurry up as rapidly as possible." It took an hour for the Federal infantry to arrive, and by then the Confederates had withdrawn in the gathering darkness.

From its bivouac near Lynnville, Wood's IV Corps had marched 16 miles on Christmas Day, closely following the cavalry, moving through Pulaski. They camped for the night six miles south of town on the Lamb's Ferry Road. Wood described the last six miles as "a road next to impracticable from the depth of the mud." He then ordered all artillery except for four batteries to remain at Pulaski. He used the surplus horses to increase the artillery teams to eight each and the caisson teams to ten each. He also limited the number of ammo wagons and reduced their cargo to ten boxes each. Otherwise, he reported, due to the miserable roads "a vigorous pursuit would have been impossible."

Forrest also attended to logistics. When Forrest had reported to Hood back in November he was ordered to reduce the number of mules per wagon and hand over the surplus to Hood's transportation quartermaster. Forrest refused and told Major A.L. Landis (according to Morton) to forget about it or else "I'll come down to his (quartermaster's) office, tie his long legs into a double bowknot around his neck, and choke him to death with his own shins … If he knew the road from here to Pulaski the order would be countermanded."

Forrest's prized mules proved critical to the Confederate retreat from Columbia onward, especially below Pulaski. Morton said, "From Richland Creek to the Tennessee River the road was strewn with (Hood's) abandoned wagons, and but for the help afforded the pontoon train by General Forrest's fine six-mule teams, great delay and probable disaster to the army would have occurred before a passage of the river was effected." After the crisis, Forrest reported to Hood that his horses had been "thoroughly broken-down."

On the topic of hardship, Federal cavalry leader Wilson, whose own horse Waif had been disabled, wrote: "The road from Pulaski to Bainbridge was as bad as it could possibly be, the country through which it runs almost entirely denuded of forage and army supplies…Men and horses suffered all the rigors of winter, snow, rain, frost, mud, and exposure. During the nights, the temperature would fall so as to make ice from half an inch to an inch thick, and this was far too thin to carry horses without breaking through. As a consequence, the roads were worked up into a continuous quagmire. The horses' legs were covered with mud, and this, in turn, was frozen, so that great numbers of the poor animals were entirely disabled, their hoofs softened and the hair of their legs so rubbed off that it was impossible for them to travel. Hundreds lost their hoofs entirely, and in all my experience I have never seen so much suffering… During the fortnight from Nashville to the Tennessee, over five thousand horses were so disabled and so worn down by fatigue, exposure, and starvation that such of them as it was not merciful to kill had to be gathered up and sent back for treatment."

## Finale
*Monday-Thursday, December 26th-29th, 1864*

Daybreak on Monday, December 26th witnessed the occurrence of two significant events — the completion of the bridge of 80 pontoon-boats across the Tennessee River and the ambush of Wilson's cavalry by Forrest at Sugar Creek, about 30 miles back.

At sunrise, the first wagons of Hood's army began the precarious trek across the narrow floating structure with General Frank Cheatham personally supervising. Due to the current, the bridge bowed slightly in the middle. The soldiers walked slowly and carefully over the unsteady structure. During the course of the day, most of the men in Cheatham's and Stevenson's (Lee's) Corps would make the crossing to safety.

Some of Hood's wagons were stalled at Pinhook Town, just past Sugar Creek. The pontoon wagons had been double-teamed, and Forrest needed to stall the Federal advance so that the horses could be brought back and the wagons pulled to Bainbridge. Forrest determined that the Sugar Creek crossing, 16 miles northeast of Lexington, was the perfect spot for another trap. The site was at a horseshoe bend in Sugar Creek where the road crossed. The dense morning fog concealed the Confederate forces almost totally. Part of the infantry was placed in front of the creek with the remainder behind the creek "to guard against disaster."

Morton's artillery was placed on a rise, aimed at the left flank of the advancing Yankee column. A detachment of Ross' cavalry supported the artillerymen. The rest of Ross' brigade backed the two units of rebel infantry 200 yards southwest of the creek — Reynolds on the left and Feild on the right. They erected field fortifications and hid behind them. The infantry units of Featherstone and Palmer were positioned a half-mile in the rear, behind a second nearby crossing of the west fork of the creek. Armstrong's cavalry covered the left

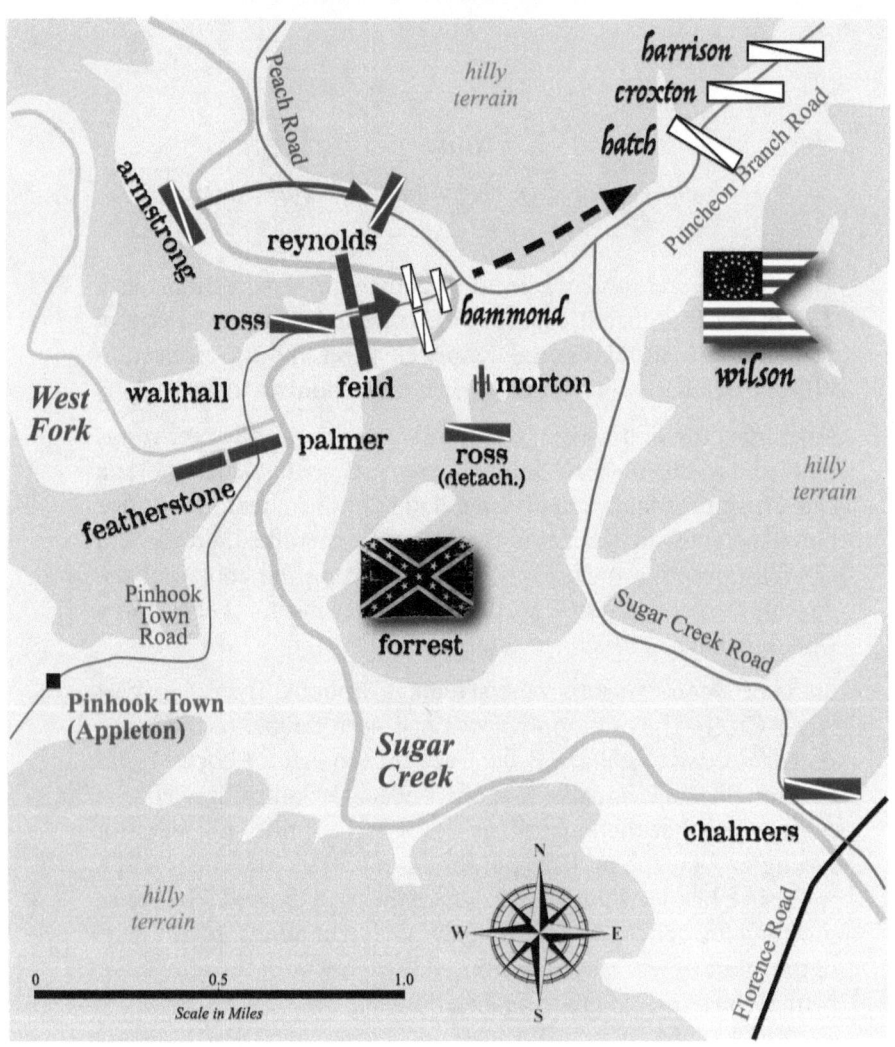

flank, while Chalmers remained a distance away on the right flank at another road crossing of the creek. While the fog diminished visibility, the gurgling of the cold creek masked any ambient noise from the troopers lying in wait.

Hammond was the designated Federal cavalry leader that day. Between 8:30 and 9:00 am, Hammond's troopers crossed the creek, dismounted and deployed into position, alert to any danger. At this point, Morton could actually see the rear of the Federal van. When the Federal troopers crept to within 30 paces, the Confederate infantry fired and then charged. Hammond's men turned and ran "in the wildest confusion," crossing the creek, where they were wracked by Morton's guns.

Walthall stated: "His flight being obstructed by the creek, we captured nearly all the horses of a dismounted regiment and some prisoners."

Supporting Walthall's infantry was Ross' brigade, which included the local Lawrence County men in Col. George H. Nixon's regiment. As they charged forward, the 2nd Mississippi of Armstrong's command crossed the creek upstream and hit the Yankees on their right flank. Just as at Anthony's Hill, the Federal troopers were driven back about a half mile into the rest of the Federal cavalry. Soon after, Forrest recalled his men and sent the infantry back on the road to Lexington. He maintained his position until noon and then fell back, although some accounts have the cavalry staying longer. Once again, outnumbered, Forrest was concerned about being outflanked.

At Sugar Creek, Forrest's men killed or wounded 150 Federal troopers and captured a dozen. Reportedly, 400 Federal horses were killed and 350 others captured.

The Federal leader Hammond described a see-saw battle: "A spirited action followed, in which the Second Tennessee, supported by the Fourth, drove the enemy into his works. A charge was made in turn by two columns of infantry, with cavalry in the center, driving us back about 300 yards across the creek, where we rallied and drove them back to their works, holding the position until the afternoon, when the Fourteenth Ohio Battery shelled their rear guard out of log-works commanding the road."

Forrest's version: "On the morning of the 26th the enemy

commenced advancing, driving back General Ross' pickets. Owing to the dense fog, he could not see the temporary fortifications which the infantry had thrown up and behind which they were secreted. The enemy therefore advanced to within 50 paces of these works when a volley was opened upon him, causing the wildest confusion. Two mounted regiments of Ross' brigade and Ector's and Granbury's brigades of infantry were ordered to charge upon the discomfited foe, which was done, producing a complete rout. The enemy was pursued for two miles, but (the enemy) showing no disposition to give battle my troops were ordered back."

1st Lieutenant J.E. Tunnell, company commander in the 14th Texas Infantry, said that after Sugar Creek, "Our brigade (Ector's) got a good Yankee breakfast from the saddle pockets on horses killed and captured. From thence to the pontoon bridge on the Tennessee R. our brigade was largely mounted."

Low on ammunition and rations, his men and horses worn, and staggering somewhat from two serious surprise attacks, Wilson unofficially gave up the pursuit of Hood's army at this point. That night he moved to Pinhook Town (now Appleton) and halted. He did send trusty Colonel George Spalding and a "flying battalion" of 500 handpicked men on the sturdiest horses to chase down Forrest the next day, but by then Forrest was too far ahead. Spalding's men didn't reach the river until the morning of December 28th, and by then it was too late.

The last opportunity to prevent Hood's army from crossing the Tennessee River rested with the U.S. Navy under Admiral Samuel Lee. However, the river level was now falling and no river pilot was willing to navigate the shoals near Bainbridge. According to the experienced pilot, the river monitor *USS Neosho*, which drew only five feet of water despite being 180 feet long, could not be deployed upriver. "The swift and shallow waters of Little Muscle Shoals were too rocky, too uneven, and too dangerous," according to naval expert and author Myron Smith Jr. Wilson later claimed that Admiral Lee had told him that he did not have a pilot trustworthy enough to navigate the tricky waters. "This was indubitably our last and best chance, but the independence of the navy and the natural timidity of a deep-water sailor in a shoal-water river defeated it."

Admiral Lee explained that, if not for the opposition of the river pilots, he "could have succeeded in reaching Bainbridge with an effective force, capable of destroying Hood's pontoons." Such pronouncements are, of course, easier to make after the fact, untested.

After covering the crossing of Cheatham's and Stevenson's Corps, A.P. Stewart's Corps crossed the river on the precarious pontoon bridge all day on Tuesday, December 27th. The pontoon bridge bulged downstream in the swift current of the wide river, and infantry had to cross single file three paces apart. After dark, Forrest and his cavalry crossed, leaving only Walthall's rearguard on the north side of the river. On December 28th at 3:00 am, Walthall set into motion so as to reach the bridge at daylight. One brigade would depart at a time, first Featherstone, then Feild and Palmer. Reynolds withdrew his command from Shoal Creek in time to reach the main line by daybreak, leaving a skirmish line behind for a half hour. Ector's brigade (consisting of 265 men, barely a third of the 700 it had at Nashville) would be the last to cross, and Coleman's 39th North Carolina the last regiment, before engineers began dismantling the bridge. Walthall directed that 200 of his men assist the engineers. By the late morning of December 28th, the Army of Tennessee had crossed and the pontoon bridge taken up.

Artilleryman Pvt. Stephenson noted: "The passage of the river being effected, we at last breathed free. The long, dreary, shameful flight was done. The control of the army appeared to be in the hands of Lieutenant General A.P. Stewart. He seemed to be our real leader throughout the retreat, and directed the passage of the river."

On December 26th, Wood's IV Infantry Corps remained in Pulaski, waiting for more rations. It appeared that the Federal infantry would fail to close with the retreating Confederates. The next day, the 1st Division of Wood's IV Corps moved at daylight, and bivouacked near Puncheon Church, on Sugar Creek. Wood's IV Corps would advance to six miles south of Lexington by the time the pursuit was called off. The pursuit, at this point, was left solely to Wilson's troopers. Wood and Wilson marked time until Thomas officially ended the pursuit on December 29th. At that point, Schofield was at Columbia, A.J. Smith was at Pulaski, Wood at Lexington, Wilson at Sugar Creek, Spalding at the river, and Admiral Lee slightly downstream.

Wood on further action: "To continue (the pursuit) farther...even, if possible, was really impossible and useless...Of the pursuit it may be truly remarked that it is without a parallel in this war. It was continued for more than a hundred miles at the most inclement season of the year, over a road the whole of which was bad, and thirty miles of which were wretched, almost beyond description. It would scarcely be hyperbole to say that the road from Pulaski to Lexington was bottomless when we passed over it. It was strewn with the wrecks of wagons, artillery carriages, and other material abandoned by the enemy in his flight."

On Thursday, December 29th, General Thomas issued General Orders No. 169 announcing the end of the campaign. "The impassable state of the roads and the consequent impossibility to supply the army compels a closing of the campaign for the present." The enemy "must forever relinquish all hope of bringing Tennessee again within the lines of the accursed rebellion."

Thomas said Hood's army "had become a disheartened and disorganized rabble of half-naked and barefooted men, who sought every opportunity to fall out by the wayside and desert their cause to put an end to their sufferings. The rear guard, however, was undaunted and firm and did its work bravely to the last."

On Jan. 18th, 1865, U.S. Grant told Sherman: "He (Thomas) is possessed of excellent judgment, great coolness, and honesty, but he is not good on a pursuit." He also said, "His pursuit of Hood indicated a sluggishness that satisfied me that he would never do to conduct one of your campaigns."

Federal surgeon George E. Cooper opined, "Probably in no part of the war have the men suffered more than in the month of December 1864."

Forrest's command suffered severely. He was faced with an enemy twice his strength, mounted on fresher horses and armed with repeating rifles, operating under the worst weather and road conditions. Retired Brig. Gen. Scales said of Forrest: "With an infantry commander he trusted and troops he inspired by personal example, he was able to accomplish an arduous mission, arguably one of the most difficult anyone could be assigned."

Military historian Jac Weller said of Forrest's performance: "He was advising Hood, not only on the rearguard action and the trains, but also on every other detail of army operation. His was a truly remarkable performance." He supervised "one of the most masterly exhibitions of military efficiency ever given by anyone at any time."

Forrest was so appalled at the condition of the army and his men he wrote in a letter to his superior on Jan. 2, 1865: "The Army of Tennessee was badly defeated and is greatly demoralized, and to save it during the retreat from Nashville I was compelled almost to sacrifice my command."

Walthall said of the retreat: "The weather was excessively severe, and the troops subjected to unusual hardships...borne without complaint."

The Army of Tennessee, numbering 18,742 men, reached Tupelo, Mississippi on Jan. 23, 1865 (Federal statistics indicate 14,500). That same day, General Richard Taylor arrived to assume command, General Hood having submitted his letter of resignation to President Jefferson Davis ten days earlier. Some of the Confederate soldiers ended the war fighting at Mobile Bay while others surrendered with Joseph Johnston in North Carolina.

"No army was ever subjected to a more severe test of fidelity and loyalty to a cause than was this army," wrote historian Hay. "Its best officers and men killed or disabled, the men in the ranks could not see how good could come from further fighting. Yet they carried on to the bitter end, proudly and bravely..."

Hay speculated: "(Hood) and Davis staked all on a desperate last move and lost. But for the long delay at Tuscumbia, Hood might have been, at least temporarily, successful..."

During 1864, the Army of Tennessee had been badly wounded at Atlanta, stunned at Franklin, routed at Nashville, yet managed somehow to survive. Hood blamed much of the failure of the campaign on his own officers and men. Hay added: "Hood was continually questioning their morale and willingness to carry on, something that no other commander had ever done before him."

Hay said: "After Franklin many of the rank and file realized the utter hopelessness of the struggle, but, as soldiers, they were bound to fight on to the bitter end until stopped by the civil government or until this government had ceased to exist."

Federal cavalry leader Wilson gave his losses for the entire Nashville campaign as 122 killed, 521 wounded, and 259 missing. Hatch's division alone captured 20 guns, 1,000 prisoners, a large number of wagons, ambulances, caissons, and gun carriages, 2 battle-flags (divisional colors), and 4 battle-flags taken with prisoners by Colonel Spalding on the second day of battle. It is estimated that at least 17 pieces of artillery were captured during the retreat from Nashville.

During December 1864, the U.S. cavalry took 3,232 prisoners, four stands of colors, 32 pieces of artillery, 2,386 stands of small arms, 184 wagons, 1,348 mules, 11 caissons, three locomotives, two hand-cars, eight ambulances, 125 pontoon wagons, and four sabers.

Wood reported capturing 25 pieces of artillery and 1,968 prisoners. His corps sustained 133 killed and 814 wounded.

Thomas in his report gave Colonel W.J. Palmer, commanding the 15th Pennsylvania Cavalry, the "credit of giving Hood's army the last blow of the campaign, at a distance of over 200 miles from where we first struck the enemy on the 15th of December, near Nashville." On December 29th, Palmer burned and destroyed the entire Confederate pontoon train at Russellville, Alabama, a train that extended for five miles and consisted of 78 pontoon-boats and about 200 wagons. Two days later, Palmer caught a supply train just across the Mississippi line, consisting of 110 wagons and 500 mules. He burned the wagons, and shot or sabered all the mules he could not lead off or use as mounts for prisoners.

The lasting effect of the Tennessee campaign cannot be overstated. "On the retreat southward in late December, soldiers of the Army of Tennessee ruthlessly stripped farms and forcibly seized conscripts, just as they had done on the advance," said historian Stephen Ashe. "Hard on their heels in pursuit came General Thomas' Yankees, and they were no less voracious and destructive...To the people of Middle Tennessee, for whom the war was an all-consuming ordeal with profound social, economic, and political consequences, Nashville was less than decisive, its outcome ambiguous. The battle may have brought an end to the clash of armies in the west, but it did not bring peace to Middle Tennessee."

"The guerilla war continued, having a logic of its own, even though the military purposes were negligible," noted historian Richard

Gildrie.

Thomas Carrick of Bedford County, Tennessee was cut off from his Confederate regiment during Hood's retreat. Until the end of May 1865 he and his small band roamed four counties bushwhacking and fighting the "dirty whelps that invaded our country."

In 1865, in response to calls for continued guerilla resistance, General N.B. Forrest, as fierce a warrior as was ever seen, stated emphatically, "Any man who is in favor of a further prosecution of this war is a fit subject for a lunatic asylum, and ought to be sent there immediately."

## Aftermath

Following the Nashville campaign, the Federal army, hodgepodge to begin with, was broken up and scattered in several directions. George Thomas was assigned few active combat duties for the remainder of the war. In March 1865, Thomas was awarded the Thanks of Congress for his victory at Nashville. Thomas never forgave himself for not having a mobile force ready to cut off Hood after the battle at Nashville. The failure was a "grave error of judgment," he later told friends, adding that Hood's army "ought all to have been captured." However, Thomas never quite took personal responsibility for sending the Federal pontoon train down the wrong road.

In August 1866, the reconstructed Tennessee Legislature appropriated $1,000 to have George Dury of Washington paint a portrait of Thomas. A gold medal commemorating his victory at Nashville was ordered from Tiffany's of New York. Thomas returned to Nashville on Dec. 15th, 1866 to accept the gold medal, presented by Governor William "Parson" Brownlow. In 1869 a Democratic member of the legislature proposed to sell the portrait, which hung in the state library. The motion failed by a large majority. Thomas angrily offered to buy it for $1,000. He vowed to return the gold medal to the legislature as soon as he could get it out of deposit in New York. The next year a meeting was arranged between Thomas and Hood, both staying at the same hotel in Louisville. In Thomas' room, they conversed for an hour about Atlanta and Nashville. Later Hood stated, "Thomas is a grand man. He should have remained with us, where he would have been appreciated and loved." On March 28, 1870, Thomas fell ill and died.

After the war, Hood moved to New Orleans and became a cotton broker and president of an insurance company. He married and sired 10 children in 11 years, including three pairs of twins. He wrote a self-serving memoir, *Advance and Retreat*. He died in 1879 at age 48 of yellow fever, which also claimed his wife and eldest daughter. Fort Hood in Texas is named for him.

Nathan Bedford Forrest ruined his health and his finances during the war. He retired to Memphis and dabbled in commercial pursuits, and served as the first "grand wizard" of the Ku Klux Klan. He disbanded the group when its tactics became too violent. He died in 1877 at age 56. As of 2007, Tennessee had 32 dedicated historical markers linked to Forrest, more than all three former U.S. Presidents (Jackson, Polk, Johnson) associated with the state combined. Forrest remains a controversial figure. In 2017, the Memphis City Council voted to remove an equestrian statue of Forrest from a city park.

In late March 1865, Harry Wilson took 13,500 well-armed troopers on a 525-mile raid through Alabama and Georgia, besting Forrest at Selma (and recapturing the 12-pounder Napoleon lost at Anthony's Hill), capturing Montgomery, burning Columbus, Georgia to the ground, and assisting in the capture of Jefferson Davis. He inflicted nearly 8,000 casualties while sustaining 725. On March 30th, Wilson ordered McCook to direct Croxton to torch Alabama University at Tuscaloosa in retribution for the school's cadets who fought at the Battle at the Barricade back on December 16th. Wilson, the boy general, died in 1925 in Wilmington, N.C., having outlived all but three other Civil War generals.

Edward Hatch was named a brevet major general after Nashville. He served in the Indian wars out West, leading the 9th U.S. Cavalry, and died in 1889.

After the war, John Croxton practiced law, established the Republican newspaper *Louisville Commercial*, and was appointed U.S. Minister to Bolivia, where he died in 1874 at age 38.

Joseph Knipe served as postmaster of Harrisburg, Pennsylvania, and other government posts until his death in 1901.

Thomas J. Harrison was wounded in North Carolina in 1865. He was brevetted brigadier general of U.S. Volunteers in that year. Following the war, he settled in Tennessee, where he served as U.S. Marshal, Middle District, in 1870. He died in 1891 at age 68.

Datus Coon was brevetted brigadier general in March 1865. He served in Alabama's reconstruction legislature and was later named as U.S. Consul in Cuba. He relocated to southern California, where he died in 1893 at age 62 of an accidental gunshot wound.

John McArthur was promoted to brevet major general and returned to Chicago, but he was never able to restore his fortunes. He served as city postmaster and commissioner of Chicago public works during the Chicago Fire. He died of a stroke at home on May 15, 1906.

Lucius Hubbard, a Republican, won the Minnesota gubernatorial election in 1881. He also participated in mining and railroading as an executive. He served in the Spanish-American War as a brigadier general of volunteers. He died in 1919.

William R. Marshall served two terms as governor of Minnesota immediately after the war. He pushed the state legislature to allow blacks to vote. He died in California in 1896 at age 70.

A career army man, John Schofield served as the Military Governor of Virginia, U.S. Secretary of War, Superintendent of West Point, and Commanding General of the United States Army (1888-95). He also recommended that the U.S. establish a naval base at Pearl Harbor, where the Schofield Barracks are named in his honor. In 1892, he was awarded the Congressional Medal of Honor for service at Wilson's Creek in 1861. He died in 1906 at age 74.

A.P. Stewart and Frank Cheatham surrendered with Johnston's forces in North Carolina in 1865. After the war, Stewart became an insurance executive. He moved to Mississippi in 1874, where he served as the Chancellor of the University of Mississippi until 1886. From 1890-1908, he was the commissioner of the Chickamauga and Chattanooga National Military Park. He died in 1908 at age 86. Cheatham ran unsuccessfully for the U.S. House in 1872. He served for four years as the appointed superintendent of a Tennessee state prison, and he was appointed postmaster of Nashville in 1885-86. He died in 1886 at age 65.

Following the war, William Hicks Jackson returned to Nashville. In 1868, he married Selene Harding, and co-managed his father-in-law's estate, Belle Meade. He and his older brother raised world-renowned prize racehorses. Jackson purchased a stallion named Iroquois in 1886, the first American winner of The Derby. Many of today's Kentucky Derby winners trace their lineage back to Belle

Meade horses. Jackson died in 1903.

Tyree Bell was appointed brigadier general and surrendered with Forrest's troops in 1865. Ten years later, he moved to California, where he became a successful farmer. He died in 1902 at age 86.

Abe Buford also surrendered with Forrest's troops. After the war, he returned to his farm in Kentucky, where he became a leading breeder of thoroughbred horses. He died in 1884 at age 64.

Frank Armstrong, who was born on the frontier and fought for both sides during the Civil War, served as U.S. Indian Inspector from 1885-89, and was the Assistant Commissioner of Indian Affairs from 1893-95. He died in 1909 at age 73.

Henry Clayton resumed his law practice and served as circuit court judge. In 1886, he accepted a position as the president of the University of Alabama, a role he held until his death in Tuscaloosa in the fall of 1889.

James Holtzclaw assumed command of the garrison at Spanish Fort and led the defenses of Mobile and Montgomery against Federal forces. He resumed his legal career and became prominent in the local and state Democratic Party. He served as an associate state railroad commissioner. He was also the grand commander of the local chapter of the Freemasons. He died in 1893 at age 59.

Randall Gibson was assigned to the defense of Mobile. He inspired his troops to hold Spanish Fort, which was under siege, until the last moment, after which they escaped. Gibson was captured in Alabama in May 1865 and paroled several days later. He served as a regent of the Smithsonian Institution, and as president of the board of administrators of Tulane University. He was elected to Congress four times and served two terms as U.S. Senator. He died in 1892. Gibson Hall, Tulane's administration building, is named for him.

In 1868, David Campbell Kelley, the "Fighting Parson," took a D.D. degree from Cumberland University and served the Methodist Episcopal Church. He was one of the founders of Vanderbilt University in 1873 and served on its board of trustees from 1875-91. He ran unsuccessfully as the Prohibition Party candidate for Tennessee governor in 1890. Kelley died in Nashville in 1909.

Daniel Reynolds had a leg blown off at Bentonville, North Carolina in March 1865. He returned to his law practice after the war and served in the Arkansas state senate. He died in 1902 at age 69.

After a brief convalescence, Daniel Govan took his brigade to North Carolina, where he surrendered with Joe Johnston's army. He returned to Arkansas and became a farmer. He died in 1911 at age 81.

After the war, Edward Walthall resumed his law practice and served many years as a U.S. Senator from Mississippi. He died in 1898 at age 67.

Stephen D. Lee settled in Mississippi. He served as a state senator in 1878, and was the first president of the Agricultural and Mechanical College of Mississippi (Mississippi State University) from 1880-99. In 1895, he was the first chairman of the Vicksburg National Park Association and was instrumental in the congressional passage of the law creating the national park in 1899. He also was an active member of the United Confederate Veterans. He died in 1908 at age 74.

One-armed Edmund Rucker was imprisoned at Johnson's Island in Ohio. Forrest organized a prisoner exchange for him and Rucker was with the army again when it surrendered at Gainesville, Alabama on May 9, 1865. After the war, he returned to Memphis; in 1869 he moved to Alabama as superintendent of a railroad. Rucker relocated to Birmingham and became an industrial magnate, dealing with coal, steel, sales, and land as well as banking. Fort Rucker in Georgia is named for him.

Rucker's sword was George Spalding's prize of war and remained in his possession for 25 years. Spalding served as the mayor of Monroe, Michigan, then as a two-term Congressman. He died in 1915. The sword was eventually returned to Rucker, who died in 1924 at age 88.

Henry R. Jackson was paroled from Fort Warren, Massachusetts, on July 8, 1865. He resumed his law practice and political career, serving as minister to Mexico from 1885-86. He also was a railroad executive, banker, and president of the Georgia Historical Society. Jackson died in Savannah in 1898 at age 77.

Capt. William LeBaron Jenny, the engineer of the U.S. pontoon-boat train, became one of the nation's foremost architects and founder of the Chicago school of skyscraper architecture.

The boy general, Thomas Benton Smith, suffering from his head wound, was imprisoned at Fort Warren in Boston Harbor. He recovered enough to do some railroad work. He ran for Congress in 1870, but lost the election. His wound caused permanent damage, and Smith spent much of his remaining 47 years in an insane asylum in Nashville, emerging occasionally for army reunions and other social events. He died in 1923 at age 85, outliving his assailant at Compton's Hill by 21 years. Formed in 1861 with 880 men, the 20th Tennessee surrendered four years later with 34 men, having fought in nearly every major battle in the Western Theater.

Visiting the battlefield in 1905, former Private John Johnston discovered that the Battle at the Barricade took place "within a few hundred yards of the grave of my Great Grandfather (Major John Johnston), an old soldier of the Revolution who had moved out from Salisbury, N.C. in 1796, and settled at this place... He owned all of this ground on both sides of the pike, and the little church, and graveyard and battleground were all on part of his land."

In February 2008, a new Battle of Nashville historical marker for the Battle of the Barricade was dedicated with ceremonies on Granny White Pike at the entrance to Richland Country Club. The battle was fought nearly a mile to the south. Working with a map drawn by Private Johnston in 1905, two Civil War enthusiasts from Nashville, Jim Kay and Fowler Low, using metal detectors to plot spent shells, discovered that the cavalry battle actually began nearly a mile farther north, where the modern marker is located. Kay is a Nashville lawyer who serves as president of the Battle of Nashville Trust; Low was an industrial sales manager and executive who passed away in 2018.

*Note: In 2020, the Battle of Nashville Preservation Society changed its name to Battle of Nashville Trust.*

In 1940, hikers discovered Granny White's grave on an overgrown hillside. The Daughters of the American Revolution relocated the grave to its current location, near the site of her original tavern, and what would be the entrance to the Inns of Granny White neighborhood. The Nashville Metropolitan Historical Commission placed a marker recognizing her achievements at her grave in 1970.

In 2004, a larger-than-life bronze statue of a U.S. Colored Troops infantryman was erected at Nashville National Cemetery, where many USCT and other Federal troops are buried.

In December 1977 police investigated the disturbance of Colonel William Shy's gravesite in Franklin. A headless body found in the damaged iron coffin was determined to be a recent unidentified murder victim. Later, after much more investigation, the body was determined to be Colonel Shy himself. Shy was re-buried with a military ceremony. The grave robbers were never caught.

Long after the war, veterans of the four Minnesota regiments in the front lines of McArthur's brigades argued about who first gained the Confederate works atop Compton's Hill. The most famous depiction of the Battle of Nashville was painted by Howard Pyle in 1906, showing Minnesota troops storming across a soggy cornfield adjacent to Compton's Hill. The massive work of art hangs in the governor's reception room at the Minnesota State Capitol. A monument and statue was erected at the Nashville National Cemetery honoring the fallen soldiers from Minnesota. The state flag flies atop Compton's Hill alongside U.S. and C.S.A. flags.

In 2014, a new monument honoring the 97 Minnesota troops who died at Compton's Hill was dedicated on the hill by the Minnesota Civil War Commemoration Task Force and the Battle of Nashville Preservation Society. The inscription on the monument, written by Minnesota Private John Milton Benthall, reads:

> "... comrades who have grown as dear to us as brothers lie dotting the steep hillside, their battles ended, their warfare over. Never more will they press with us shoulder to shoulder as the bristling steel points sweep resistlessly on, never more in our hours of glee will their voices join in the merry jest or fill the air with laughter — they are gone. And the everlasting mountains in the shadow of which they lie shall be their eternal monument; year after year the forest trees will shed their crowns of glory over them, and day by day the winds, as they sigh through the Brentwood Hills, will chant a low, sad requiem to their memory."

The summit of Shy's (Compton's) Hill today. Signage from the trailhead.

Beauregard's howitzers (replicas) on the eastern slope.

The Minnesota Monument on Shy's Hill.

## Battle of Nashville
## Order of Battle

# U.S. Army
## Maj. Gen. George H. Thomas, Commanding

### Major General George Henry Thomas
### (1816-70)

A native Virginian, he was disowned by his family for remaining loyal to the Union. He graduated West Point in 1840, ranked 12th in the class. He served in the artillery during the Seminole and Mexican wars. He served in the 2nd Cavalry under Albert Sidney Johnston and Robert E. Lee. In the Civil War, he commanded troops defeating Confederates under Zollicoffer at Fishing Creek, Ky. in Jan. 1862. He was known as the "Rock of Chickamauga" for standing firm at that battle in Sept. 1863 while the rest of Rosecrans' Union army was routed. His men from the Army of the Cumberland stormed Missionary Ridge without orders and broke the seige of Chattanooga in Nov. 1863. His army was the central force in Sherman's campaign against Atlanta, May-Sept. 1864. Following the victory at Nashville, Thomas was promoted to Major General, U.S. Army, and received the "Thanks of Congress." He served in the Army until his death in California. He is buried at Oakwood Cemetery, Troy, NY.

## XVI Corps or Detachment Army of the Tennessee: Maj. Gen. Andrew J. Smith

**Major Gen. Andrew Jackson Smith** (1815-97) Born in Pennsylvania, he graduated from West Point in 1838 and served with the Dragoons (cavalry) in the West for 23 years. His troops served with Sherman at Chickasaw Bluffs and Vicksburg, Miss., and in the Red River campaigns. He defeated Forrest at Tupelo on July 14, 1864. His rugged troops were assigned to so many different locales that they became known as the "lost tribes of Israel" and "Smith's guerillas." After the war he served as postmaster and city auditor of St. Louis, Mo. He is buried at Bellefontaine Cemetery, St. Louis.

### First Division: Brig. Gen. John McArthur

1st Brigade: Col. William L. McMillen
114th Illinois; 93rd Indiana; 10th Minnesota

2nd Brigade: Col. Lucius F. Hubbard
5th, 9th Minnesota; 11th Missouri; 8th Wisconsin;
2nd Battery Iowa Light Artillery (Reed)

3rd Brigade: Col. Sylvester G. Hill (k), Col. William R. Marshall
12th, 35th Iowa; 7th Minnesota; 33rd Missouri;
Battery I 2nd Missouri Light Artillery (Julian)

### Second Division: Brig. Gen. Kenner Garrard

1st Brigade: Col. David Moore
119th, 122nd Illinois; 89th Indiana; 21st Missouri;
9th Battery Indiana Light Artillery (Calfee)

2nd Brigade: Col. James L. Gilbert
58th Illinois; 27th, 32nd Iowa; 10th Kansas;
3rd Battery Indiana Light Artillery (Ginn)

3rd Brigade: Col. Edward H. Wolfe
49th, 117th Illinois; 52nd Indiana; 178th New York;
Battery G, 2nd Illinois Light Artillery (Lowell)

### Third Division: Col. Jonathan B. Moore

1st Brigade: Col. Lyman M. Ward
72nd Illinois; 40th Missouri; 14th, 33rd Wisconsin

2nd Brigade: Col. Leander Blanden
81st, 95th Illinois; 44th Missouri Artillery;
11th Battery Indiana Light Artillery (Morse);
Battery A, 2nd Missouri Light Artillery (Zepp)

Key to Abbreviations:
k= killed;   w= wounded
mw= mortally wounded;   c= captured

## IV Army Corps: Brig. Gen. Thomas J. Wood

**Brig. Gen. Thomas John Wood** (1823-1906)
Born in Kentucky, he graduated West Point in 1845 and won honors in the Mexican War. He saw action at Shiloh under Buell; at Perryville, Ky.; and at Murfreesboro, TN, where he was wounded. In a controversial incident at Chickamauga in Sept. 1863 he moved his division under orders and opened a gap in the Union lines which allowed Longstreet to rout the Union right wing. His men were first to crest Missionary Ridge at Chattanooga in Nov. 1863. He was again wounded at Lovejoy Station, Sept. 1864. In 1865 he was promoted to Major General and retired in 1868. He died in Dayton, Ohio. He is buried at West Point, NY.

### First Division: Brig. Gen. Nathan Kimball

**1st Brigade: Col. Isaac M. Kirby**
21st, 38th Illinois; 31st, 81st Indiana; 90th Ohio

**2nd Brigade: Brig. Gen. Walter C. Whittaker**
96th, 115th Illinois; 35th Indiana; 21st, 23rd Kentucky; 45th, 51st Ohio

**3rd Brigade: Brig. Gen. William Grose**
75th, 80th, 84th Illinois; 9th, 30th, 36th, 84th Indiana; 77th Pennsylvania

### Second Division: Brig. Gen. Washington L. Elliott

**1st Brigade: Col. Emerson Opdycke**
36th, 44th, 73rd, 74th, 88th Illinois; 125th Ohio; 24th Wisconsin

**2nd Brigade: Col. John Q. Lane**
100th Illinois; 40th, 57th Indiana; 28th Kentucky; 26th, 97th Ohio

**3rd Brigade: Col. Joseph Conrad**
42nd, 51st, 79th Illinois; 15th Missouri; 64th, 65th Ohio

### Third Division: Brig. Gen. Samuel Beatty

**1st Brigade: Col. Abel D. Streight**
89th Illinois; 51st Indiana; 8th Kansas; 15th, 49th Ohio

**2nd Brigade: Col. P. Sidney Post (w)**
59th Illinois; 41st, 71st, 93rd, 124th Ohio

**3rd Brigade: Col. Frederick Knefler**
79th, 86th Indiana; 13th, 19th Ohio

**Artillery: Maj. Wilbur F. Goodspeed**
Light Batteries: 25th Indiana (Sturm); 1st Kentucky (Thomason); 1st Michigan (De Vries); 1st Ohio G (Marshall); 6th Ohio (Baldwin); Battery B, Pennsylvania Light Artillery (Ziegler); Battery M, 4th U.S. (Canby)

## XXIII Army Corps: Maj. Gen. John M. Schofield

Major Gen. John McAllister Schofield (1831-1906) Born in New York and raised in the Midwest, he graduated from West Point in 1853, ranking 7th in the class. He commanded militia and the Army of the Frontier in Missouri from 1861-63. He led the XXIII Corps during Sherman's Atlanta campaign and inflicted heavy losses on Hood's army at the Battle of Franklin, Nov. 30, 1864. After the war he served as Secretary of War under President Johnson, West Point superintendent (1876-81), and commander of the U.S. Army in 1888-95.

He recommended that Pearl Harbor be acquired as a naval base. He is buried at Arlington National Cemetery.

| Second Division: Maj. Gen. Darius N. Couch |
|---|
| 1st Brigade: Brig. Gen. Joseph A. Cooper<br>130th Indiana; 26th Kentucky; 25th Michigan; 99th Ohio;<br>3rd, 6th Tennessee |
| 2nd Brigade: Col. Orlando H. Moore<br>107th Illinois; 80th, 129th Indiana; 23rd Michigan; 111th, 118th Ohio |
| 3rd Brigade: Col. John Mehringer<br>91st, 123rd Indiana; 50th, 183rd Ohio |
| Artillery:<br>Light Batteries: 13th Indiana (Harvey); 19th Ohio (Wilson) |

| Third Division: Brig. Gen. Jacob D. Cox |
|---|
| 1st Brigade: Col. Charles C. Doolittle<br>12th, 16th Kentucky; 100th, 104th Ohio; 8th Tennessee |
| 2nd Brigade: Col. John S. Casement<br>65th Illinois; 65th, 124th Indiana; 103rd Ohio; 5th Tennessee |
| 3rd Brigade: Col. Israel N. Stiles<br>112th Illinois; 63rd, 120th, 128th Indiana |
| Artillery:<br>Light Batteries: 23rd Indiana (Wilber); Battery D, 1st Ohio (Cockerill) |

## Provisional Detachment, District of the Etowah: Maj. Gen. James B. Steedman

**Major Gen. James Blair Steedman (1817-83)**
A Pennsylvanian, he was a civilian general who had little formal education and had worked as a printer, Ohio legislator, and owner of the *Toledo Times* newspaper. Active in Democratic Party politics. His men served at Perryville, Ky.; Murfreesboro, Tenn.; and at Chickamauga, where he led an attack carrying the regimental colors after his horse was shot from under him. He commanded the Post of Chattanooga from Oct. 1863 to May 1864. After the war he collected revenues in New Orleans, and then edited a paper in Toledo and served as chief of police there. He is buried at Woodlawn Cemetery, Toledo, Ohio.

**Provisional Division: Brig. Gen. Charles Cruft**

1st Colored Brigade: Col. Thomas J. Morgan
14th, 16th, 17th, 18th, 44th U.S. Colored Troops

2nd Colored Brigade: Col. Charles R. Thompson
12th, 13th, 100th U.S. Colored Troops;
1st Battery Kansas Light Artillery (Tennessee)

1st Brigade: Col. Benjamin Harrison
3 battalions from 20th Army Corps

2nd Brigade: Col. John C. Mitchell
men from detached duty Army of the Tennessee

3rd Brigade: Lt. Col. Charles H. Grosvenor
68th Indiana; 18th, 121st Ohio; 2nd Battalion 14th Army Corps

Artillery:
20th Battery Indiana Light Artillery (Osborne);
18th Battery Ohio Light Artillery (Aleshire)

## Cavalry Corps: Maj. Gen. James H. Wilson

**Major Gen. James Harrison Wilson (1837-1925)**
Born in Illinois, he graduated West Point in 1860, sixth in his class. He was a topographical engineer and an aide to McClellan and Grant. He was inspector general of the Army of the Tennessee during the Vicksburg campaign. In Oct. 1863 he was promoted to brigadier general of volunteer infantry. After Chattanooga and Knoxville, he was named chief of the cavalry bureau in Washington. He commanded a division of Sheridan's cavalry in Virginia and in Oct. 1864 he became chief of cavalry of Sherman's Military Division of the Mississippi.  In 1865 he led one of the war's most successful cavalry raids into Alabama and Georgia, defeating Forrest in the process. Retiring in 1870, he managed railroads, traveled, and wrote on a variety of topics. At the age of 61, he volunteered and fought in the Spanish-American War in Puerto Rico and Cuba and in the Boxer Rebellion in China. He is buried at Old Swedes Churchyard, Wilmington, N.C.

### First Division: Brig. Gen. Edward M. McCook

**1st Brigade: Brig. Gen. John T. Croxton**
8th Iowa; 4th Kentucky Mounted Infantry; 2nd Michigan; 1st Tennessee; Board of Trade Battery, Illinois Light Artillery (Robinson)

**2nd Brigade: Col. Oscar H. La Grange**
Detached in pursuit of Lyon's raid into western Kentucky

**3rd Brigade: Bvt. Brig. Gen. Louis D. Watkins**
Detached in pursuit of Lyon's raid into western Kentucky

### Fifth Division: Brig. Gen. Edward Hatch

**1st Brigade: Col. Robert R. Stewart**
3rd Illinois; 11th Indiana; 12th Missouri; 10th Tennessee

**2nd Brigade: Col. Datus E. Coon**
6th, 7th, 9th Illinois; 2nd Iowa; 12th Tennessee; Battery I, 1st Illinois Light Artillery (McCartney)

### Sixth Division: Brig. Gen. Richard W. Johnson

**1st Brigade: Col. Thomas J. Harrison**
16th Illinois; 5th Iowa; 7th Ohio

**2nd Brigade: Col. James Biddle**
14th Illinois; 6th Indiana; 8th Michigan; 3rd Tennessee
Artillery: Battery I, 4th U.S. (Frank G. Smith)

### Seventh Division: Brig. Gen. Joseph F. Knipe

**1st Brigade: Bvt. Brig. Gen. John H. Hammond**
9th, 10th Indiana; 19th Pennsylvania; 2nd, 4th Tennessee

**2nd Brigade: Col. Gilbert M.L. Johnson**
12th, 13th Indiana; 8th Tennessee Artillery; 14th Battery Ohio Light Artillery (Myers)

## Battle of Nashville
## Order of Battle

# Confederate Army of Tennessee
### Maj. Gen. John Bell Hood, Commanding

### General John Bell Hood (1831-79)

Born in Kentucky and graduated from West Point in 1853, he gained fame as commander of the Texas Brigade at Gaines Mill, Va. and rapidly advanced in rank throughout the war. He fought at Second Manassas and Antietam and then led a division under Longstreet at Gettysburg, where his left arm was crippled. At Chickamauga in Georgia, he was badly wounded and his right leg amputated. Convalescing in Richmond, he cultivated his friendship with President Davis. A Corps commander in the Atlanta campaign, he succeeded Johnston as commander of the Army of Tennessee in July 1864. As was his nature, he immediately took the offensive, suffered consecutive defeats, and eventually surrendered Atlanta in Sept. 1864. Confounded at Spring Hill, Tenn. during his 1864 Tennessee offensive, he unwisely attacked Union fortifications at Franklin on Nov. 30, 1864 and suffered heavy losses. After the defeat at Nashville, he was relieved of command in Jan. 1865. After the war, he became a merchant and married, having 11 children in 10 years. He, his wife, and one child died of yellow fever in the epidemic of 1878. He is buried at Metairie Cemetery in New Orleans, La.

## Stewart's Corps: Lt. Gen. Alexander P. Stewart

Lt. Gen. Alexander Peter Stewart (1821-1908) Born in east Tennessee, he graduated West Point in 1842 and resigned in 1845 to become the chair of mathematics and philosophy at Cumberland College in Lebanon, Tenn. and at Nashville University. He was an anti-secessionist Whig who volunteered to fight for the South. Known as "Old Straight," he led a brigade in Polk's command in all battles of the Army of Tennessee until June 1864, when he assumed corps command upon Polk's death. He surrendered with Johnston in North Carolina in 1865. A businessman and educator, he served as Chancellor of the University of Mississippi from 1874-86 and was instrumental in establishing the Chattanooga-Chickamauga National Military Park. He is buried at St. Louis, Mo.

### Loring's Division: Maj. Gen. William W. Loring

Featherston's Brigade: Brig. Gen. Winfield S. Featherston
1st, 3rd, 22nd, 31st, 33rd, 40th Mississippi; 1st Miss. Battalion

Adams's Brigade: Col. Robert Lowry
6th, 14th, 15th, 20th, 23rd, 43rd Mississippi

Scott's Brigade: Col. John Snodgrass
27th, 35th, 49th, 55th, 57th Alabama; 12th Louisiana

### French's Division: Brig. Gen. Claudius Sears (w, c)

Ector's Brigade: Col. David Coleman
29th, 30th North Carolina; 9th Texas; 10th, 14th, 32nd Texas Cavalry

Cockrell's Brigade: Col. Peter C. Flournoy
1st, 2nd, 3rd, 4th, 5th, 6th Missouri; 1st Missouri Cavalry;
3rd Missouri Cavalry Battalion

Sears's Brigade: Lt. Col. Reuben H. Shotwell
4th, 35th, 36th, 39th, 46th Mississippi; 7th Mississippi Battalion

### Walthall's Division: Maj. Gen. Edward C. Walthall

Quarles's Brigade: Brig. Gen. George D. Johnston
1st Alabama; 42nd, 46th, 48th, 49th, 53rd, 55th Tennessee

Cantey's Brigade: Brig. Gen. Charles M. Shelley
17th, 26th, 29th Alabama; 37th Mississippi

Reynolds's Brigade: Brig. Gen. Daniel H. Reynolds
4th, 9th, 25th Arkansas; 1st, 2nd Arkansas Mounted Rifles

## Lee's Corps: Lt. Gen. Stephen D. Lee

**Lt. Gen. Stephen Dill Lee** (1833-1908) A native of South Carolina, he graduated from West Point in 1833. He served as an artillerist in the Eastern Theater through Sharpsburg and then commanded the Confederate artillery at Vicksburg, Miss. Captured there in July 1863, he was released months later and placed in command of cavalry in part of the Western Theater. The youngest lieutenant general in the CSA, he assumed command of Hood's Corps when Hood lead the Army of Tennessee. He served with Johnston in North Carolina when the war ended. He lived in Mississippi as a farmer, state senator, and the first president of Mississippi State University. He was a leading figure in the United Confederate Veterans. He is buried in Columbus, Miss.

**Johnson's Division: Maj. Gen. Edward Johnson (c)**

Deas's Brigade: Brig. Gen. Zachariah C. Deas
19th, 22nd, 25th, 38th, 50th Alabama

Manigault's Brigade: Lt. Col. William L. Butler
24th, 28th, 34th Alabama; 10th, 19th South Carolina

Sharp's Brigade: Brig. Gen. Jacob H. Sharp
7th, 9th, 10th, 41st, 44th Mississippi; 9th Battalion Miss. Sharpshooters

Brantley's Brigade: Brig. Gen. William F. Brantley
24th, 27th, 29th, 30th, 34th Mississippi; Dismounted Cavalry Company

**Stevenson's Division: Maj. Gen. Carter L. Stevenson**

Cummings's Brigade: Col. Elihu P. Watkins
24th, 36th, 39th, 56th Georgia

Pettus's Brigade: Brig. Gen. Edmund W. Pettus
20th, 23rd, 30th, 31st, 46th Alabama

**Clayton's Division: Maj. Gen. Henry D. Clayton**

Stovall's Brigade: Brig. Gen. Marcellus A. Stovall
40th, 41st, 42nd, 43rd, 52nd Georgia

Gibson's Brigade: Brig. Gen. Randall L. Gibson
1st, 4th, 13th, 16th, 19th, 20th, 25th, 30th Louisiana;
4th Louisiana Battalion; 14th Louisiana Battalion Sharpshooters

Holtzclaw's Brigade: Brig. Gen. James Holtzclaw
18th, 32nd, 36th, 38th, 58th Alabama

### Cheatham's Corps: Maj. Gen. Benjamin F. Cheatham

Major Gen. Benjamin Franklin Cheatham (1820-86)
A native Nashvillian, he served in the Mexican War as Colonel of the Tennessee Volunteers. While a farmer, he also served as Major General of the state militia. He commanded a brigade, division, then corps in the Army of Tennessee in every battle from Shiloh to Atlanta. Cheatham assumed command of Hardee's Corps as Hood began his 1864 Tennessee offensive. Blamed by Hood for the Union escape at Spring Hill, Tenn., Cheatham's Corps suffered heavily attacking fortifications the next day at Franklin. After the war, he ran unsuccessfully for Congress, and served as superintendent of state prisons and postmaster of Nashville. A hard fighter and hard drinker, he was universally beloved by his men. He is buried at Mt. Olivet Cemetery, Nashville.

#### Cleburne's Division: Brig. Gen. James A. Smith

Govan's Brigade: Brig. Gen. Daniel C. Govan
1st, 2nd, 5th, 6th, 7th, 8th, 13th, 15th, 19th, 24th Arkansas

Lowrey's Brigade: 16th, 33rd, 45th Alabama;
5th, 8th, 32nd Mississippi; 3rd Miss. Battalion

Granbury's Brigade: Capt. E.T. Broughton
5th Confederate; 35th Tennessee; 6th, 7th, 10th, 15th Texas;
17th, 18th, 24th, 25th Texas Cavalry; Nutt's Louisiana Cavalry

Smith's Brigade: Col. Charles H. Olmstead
54th, 57th, 63rd Georgia; 1st Georgia Volunteers

#### Brown's Division: Brig. Gen. Mark P. Lowrey

Gist's Brigade: Lt. Col. Zachariah L. Watters
46th, 65th Georgia; 2nd Battalion Ga. Sharpshooters;
16th, 24th South Carolina

Maney's Brigade: Col. Hume R. Feild
1st, 4th, 6th, 8th, 9th, 16th, 27th, 28th, 50th Tennessee

Strahl's Brigade: Col. Andrew J. Kellar
4th, 5th, 19th, 24th, 31st, 33rd, 38th, 41st Tennessee

Vaughan's Brigade: Col. William M. Watkins
11th, 12th, 13th, 29th, 47th, 51st, 52nd, 154th Tennessee

#### Bate's Division: Maj. Gen. William B. Bate

Tyler's Brigade: Brig. Gen. Thomas Benton Smith (w, c)
37th Georgia; 4th Battalion Ga. Sharpshooters;
2nd, 10th, 20th, 37th Tennessee

Finley's Brigade: Maj. Glover Ball
1st, 3rd, 4th, 6th, 7th Florida; 1st Florida Cavalry

Jackson's Brigade: Brig. Gen. Henry R. Jackson (c)
25th, 29th, 30th Georgia; 1st Ga. Confederate;
1st Battalion Georgia Sharpshooters

## CAVALRY
## Maj. Gen. Nathan Bedford Forrest

Major Gen. Nathan Bedford Forrest (1821-77)

Born in Bedford County, Tenn., he was a self-taught man, rising from poverty to millionaire planter and slave dealer in Memphis. He enlisted as a private in 1861 and equipped his own battalion of cavalry. He led his men out of Fort Donelson before its capture and served at Shiloh. He was promoted to brigadier general after he captured the garrison at Murfreesboro, Tenn. in July 1862. Known as the "Wizard of the Saddle," his cavalry exploits in the Union rear of the Western Theater caused much frustration on the part of U.S. Army commanders. His most brilliant victory was Brice's Crossroads, Miss.; his most controversial the capture of Fort Pillow, Tenn., known in the North as the Fort Pillow Massacre. His cavalry command destroyed the large Union supply depot at Johnsonville, Tenn. in Nov. 1864. Often contentious with superiors, he nevertheless commanded Hood's cavalry during the 1864 Tennessee offensive. He was detached to Murfreesboro during the Battle of Nashville and was stymied at the Battle of the Cedars. After the war, he again became a planter and managed a railroad. He was affiliated with the original Ku Klux Klan but resigned when it became too violent. He is buried at Forrest Park in Memphis, Tenn.

| Chalmers's Division: Brig. Gen. James R. Chalmers |
|---|
| Rucker's Brigade: Col. Edmund W. Rucker (w, c)<br>7th Alabama; 5th Mississippi; 7th, 12th, 14th, 15th Tennessee;<br>Forrest's Regiment Tennessee Cavalry |
| Biffle's Brigade: Col. Jacob B. Biffle<br>10th Tennessee |
| **Buford's Division: Brig. Gen. Abraham Buford** |
| Bell's Brigade: Col. Tyree H. Bell<br>2nd, 19th, 20th, 21st Tennessee; Nixon's Tennessee Regiment |
| Crossland's Brigade: Col. Edward Crossland<br>3rd, 7th, 8th, 12th Kentucky Mounted Infantry; 12th Kentucky;<br>Huey's Kentucky Battalion |
| **Jackson's Division: Brig. Gen. William H. Jackson** |
| Armstrong's Brigade: Brig. Gen. Frank C. Armstrong<br>1st, 2nd, 28th Mississippi; Ballentine's Mississippi Regiment |
| Ross's Brigade: Brig. Gen. Lawrence S. Ross<br>5th, 6th, 9th Texas; 1st Texas Legion |
| Artillery: Morton's Tennessee Battery |

## Confederate Artillery

### Lee's Corps: Maj. John W. Johnston

Courtney's Battalion: Capt. James P. Douglas
Dent's Alabama Battery; Douglas's Texas Battery;
Garritty's Alabama Battery

Eldridge's Battalion: Capt. Charles E. Fenner
Eufaula Alabama Battery; Fenner's Louisiana Battery;
Stanford's Mississippi Battery

Johnson's Battalion: Capt. John B. Rowan
Corput's Georgia Battery; Marshall's Tennessee Battery;
Stephens's Light Artillery

### Stewart's Corps: Lt. Col. Samuel C. Williams

Trueheart's Battalion:
Lumsden's Alabama Battery; Selden's Alabama Battery

Myrick's Battalion:
Bouanchaud's Louisiana Battery; Cowan's Mississippi Battery;
Darden's Mississippi Battery

Storrs's Battalion:
Guiborps Missouri Battery; Hoskin's Mississippi Battery;
Kolb's Alabama Battery

### Cheatham's Corps: Col. Melancthon Smith

Hoxton's Battalion:
Perry's Florida Battery; Phelan's Alabama Battery;
Turner's Mississippi Battery

Hotchkiss's Battalion:
Bledsoe's Missouri Battery; Goldtwaite's Alabama Battery;
Key's Arkansas Battery

Cobb's Battalion:
Ferguson's South Carolina Battery; Phillip's Tennessee Battery;
Slocumb's Louisiana Battery

# BIBLIOGRAPHY

Allyn, John, "Minnesota Troops at the Battle of Nashville," Battle of Nashville Trust website, www.bonps.org.

American Battlefield Trust website, www.battlefields.org/learn/civil-war/battles/nashville

Ash, Stephen V., "The Aftermath: Middle Tennessee, December 1864 to May 1865," Battle of Nashville 140th Anniversary Symposium, Nashville, Tennessee, December 2004.

Battle of Nashville Trust website, www.bonps.org, various citations. Also battlesite signage.

Boynton, Henry V., *Was General Thomas Slow at Nashville?*, New York, Harper, 1896.

Bradley, Michael R., *Forrest's Fighting Preacher*, The History Press, 2011.

——————, *Nathan Bedford Forrest's Escort and Staff*, Pelican Publishing Co., Inc., 2006.

Bragg, C.L., *Distinction in Every Service, Brigadier General Marcellus A. Stovall, CSA*, White Mane Books, 2002.

Carley, Kenneth, *Minnesota in the Civil War: An Illustrated History*, Minnesota Historical Society Press, 2000.

Carlock, Chuck, *History of the Tenth Texas Cavalry (Dismounted) Regiment, 1861-1865*, Smithfield Press, 2001.

Cartwright, Thomas; Elliott, Sam Davis; McDonough, James Lee; and Sword, Wiley. Battle of Nashville Symposium at Travellers Rest, Nashville, October 2004.

Cartwright, Thomas; Currey, David; and Massey, Ross. Peach Orchard Hill Symposium at Travellers Rest, Nashville, January, 2005.

Chalmers, J.R., *Lieutenant General Nathan Bedford Forrest And His Campaigns*, Southern Historical Society Papers, Vol. VII. Richmond, Va., October 1879. No. 10.

Cooling, Benjamin Franklin, "The Decisive Battle of Nashville," Civil War Trust website, www.civilwar.org/learn/articles/decisive-battle-nashville

Cotner, James R., "America's Civil War: Horses and Field Artillery," America's Civil War magazine, March 1996. www.historynet.com/civil-war-cannon

Cox, Jacob D., *Sherman's March to the Sea: Hood's Tennessee Campaign and the Carolina Campaigns of 1865*, Da Capo Press, 1994.

Cozzens, Peter, "Irresistible Force: Wilson's Ruthless Raid Extinguished Confederate Hopes in the Southern Heartland," America's Civil War, Vol. 30, No. 3, July 2017, pages 28-37.

Daniel, Larry J., *Cannoneers in Gray: The Field Artillery of the Army of Tennessee, 1861-1865*, University of Alabama Press, 1984.

Davis, Sue (Hendrix) and Dean, Linda Jo (Hendrix), *On the Banks of Sugar Creek*, July 2003 (out of print)

# BIBLIOGRAPHY

Edwards, Elijah Evan, "Civil War Diary of Elijah Evan Edwards, Chaplain, 7th Minnesota Volunteer Infantry," Battle of Nashville Preservation Society website. Ruth P. and Roy L. Cunningham transcription, June 2011.

Elliott, Sam Davis, *Soldier of Tennessee: General Alexander P. Stewart and the Civil War in the West*, Louisiana State University Press, 1999.

Flagel, Thomas, "Impact of Forts on the Battle of Nashville," Battle of Nashville 153rd Anniversary Symposium, December 2017.

Foote, Shelby, *The Civil War: A Narrative*, Random House, 1974.

Frey, David, *Failure to Pursue: How the Escape of Defeated Forces Prolonged the Civil War*, McFarland & Company, Inc., 2016.

Groom, Winston, *Shrouds of Glory: From Atlanta to Nashville: The Last Great Campaign of the Civil War*, Atlantic Monthly Press, 1995.

Hay, Thomas Robson, *Hood's Tennessee Campaign*, Morningside Press, 1976.

Hollandsworth, James G. Jr., "The Execution of White Officers from Black Units by Confederate Forces During the Civil War," *Black Flag Over Dixie: Racial Atrocities and Reprisals in the Civil War*, ed. by Urwin, Gregory, Southern Illinois University Press, 2004.

Hood, John Bell, *Advance & Retreat*, University of Nebraska Press, 1996.

Hood, O.C., *The Army of Tennessee in Retreat: From Defeat at Nashville through the Sternest Trials of the War*, McFarland & Company, Inc., 2018.

Horn, Stanley F., *The Decisive Battle of Nashville*, Louisiana State University Press, 1956.

Hughes, Nathaniel Cheairs, Jr., ed., *The Civil War Memoir of Philip Daingerfield Stephenson, D.D.*, UCA Press, 1995.

——————————— , *Brigadier General Tyree H. Bell, CSA: Forrest's Fighting Lieutenant*, University of Tennessee Press, 2004.

Hurst, Jack, *Nathan Bedford Forrest: A Biography*, Random House, 1993.

Johansson, M. Jane, "Gibson's Louisiana Brigade During the 1864 Tennessee Campaign," Tennessee Historical Quarterly, Vol. LXIV, No. 3, Fall 2005, pages 186-195.

Johnston, John, "Personal Reminiscences of the Civil War," Battle of Nashville Trust website.

Keenan, Jerry, *Wilson's Cavalry Corps: Federal Campaigns in the Western Theater, October 1864 Through Spring 1865*, McFarland & Company, Inc., 1998.

Keller, David L., "The North's Lost POW Camp," MHQ: The Quarterly Journal of Military History, Vol. 31, No. 2.

Kemmerly, Phillip R., "Fighting and Dying in a Frozen Hell: The Impact of Ice, Snow, Fog, and Frozen-Hard Ground on the Battle of Nashville," Tennessee Historical Quarterly, Vol. LXXIV, No. 2, p. 74-103.

# BIBLIOGRAPHY

Kennedy, Edwin Lt. Col. (ret), "Horses in the War Between the States," Nashville Civil War Round Table presentation, Sept. 17, 2019.

Logsdon, David R. (ed.), *Eyewitnesses at the Battle of Nashville*, Kettle Mills Press, 2004.

——————————, *Eyewitnesses at the Battle of Franklin*, Kettle Mills Press, 2000.

Longacre, Edward G., *Grant's Cavalryman: The Life and Wars of General James H. Wilson*, Stackpole Books, 1972.

Massey, Ross, *Nashville Battlefield Guide*, Tenth Amendment Publishing, 2007.

McAulay, John D., *Civil War Breech Loading Rifles*, Andrew Mowbray Publishers, 1987.

McDonough, James Lee, *Nashville: The Western Confederacy's Final Gamble*, University of Tennessee Press, 2004.

McMurry, Richard M., *John Bell Hood and the War for Southern Independence*, University Press of Kentucky, 1982.

McRae, James S., "I Don't Want to Be in a Hotter Place": The Twelfth USCT at the Battle of Nashville, or David C. Cooke Joins the USCT," Battle of Nashville 140th Anniversary Symposium, Nashville, Tennessee, December 2004.

Mellor, A.M., "Forrest Twice Sought Walthall: A Research Note on Leadership in the Western Theater," Tennessee Historical Quarterly, Vol. LXXIV, No. 2, p.128-141.

Parfitt, Allen, "The Battle of Franklin," Military History Online.

Pittman, Walter Earl, "A Forlorn Hope: The Confederate Rearguard After Nashville," Battle of Nashville 140th Anniversary Symposium, Nashville, Tennessee, December 2004.

Rafuse, Ethan S., "Underrated: 10 Confederate Commanders You Should Get to Know," America's Civil War, Nov. 2017.

Scales, John R., *The Battles and Campaigns of Confederate General Nathan Bedford Forrest*, Savas Beatie, 2017.

Schiller, Laurence D., "The Evolution of the Federal Cavalry 1861-1865," Essential Civil War Curriculum, June 2017.

Sides, Hampton, *On Desperate Ground*, Doubleday, 2018.

Simpson, Colonel Harold B., *Cry Commanche: The 2nd U.S. Cavalry in Texas, 1855-1861*, Hill College Press, 1979.

Smith, Derek, *In the Lion's Mouth: Hood's Tragic Retreat from Nashville, 1864*, Stackpole Books, 2011.

## BIBLIOGRAPHY

Smith, Myron J. Jr., *Tinclads in the Civil War*, McFarland & Company, Inc. Publishers, 2010.

Smith, Timothy B., *Shiloh: Conquer or Perish*, University of Kansas Press, 2014.

Stockdale, Paul H., *The Death of An Army: The Battle of Nashville and Hood's Retreat*, Southern Heritage Press, 1992.

Sword, Wiley, *The Confederacy's Last Hurrah: Spring Hill, Franklin & Nashville*, University Press of Kansas, 1992.

Watkins, Sam, *Co. Aytch: A Side Show of the Big Show*, Simon & Schuster, 1990.

*War of the Rebellion: A Compilation of the Official Records of the Federal and Confederate Armies*, Govt. Printing Office, Washington, D.C., 1886.

Warner, Ezra W., *Generals in Blue*, Louisiana State University Press, 1964.

——————, *Generals in Gray*, Louisiana State University Press, 1959.

Weller, Jac, "Nathan Bedford Forrest: An Analysis of Untutored Military Genius," *Nathan Bedford Forrest and the Confederate Cavalry in Western Tennessee*, Tennessee Historical Society, 2013.

Wigginton, Thomas A., Hood's Nashville Campaign, "Cavalry Operations," Civil War Times Illustrated, 1964.

Wills, Brian Steel, *The Confederacy's Greatest Cavalryman: Nathan Bedford Forrest*, University Press of Kansas, 1992.

Woodworth, Steven E., *Jefferson Davis and His Generals*, University Press of Kansas, 1990.

Zimmerman, Mark, *Guide to Civil War Nashville-2nd Edition*, Zimco Publications LLC, 2019.

## About the Author

Mark Zimmerman is a retired newspaperman and publications manager who resides in Nashville, Tennessee. In addition to operating Zimco Publications LLC, he is a member of the American Battlefield Trust, Battle of Nashville Trust, Nashville Civil War Roundtable, Save the Franklin Battlefield, Civil War Fortification Study Group, and other historical preservation groups. He enjoys traveling to historic sites and museums, a good joke, and collecting books and old bottles of whiskey. More info available at zimcopubs.com.

www.ingramcontent.com/pod-product-compliance
Lightning Source LLC
Chambersburg PA
CBHW020651300426
44112CB00007B/332